Domestic Animal Behavior: Causes and Implications for Animal Care and Management

Domestic Animal Behavior: Causes and Implications for Animal Care and Management

JAMES V. CRAIG

Department of Animal Sciences and Industry
Kansas State University

Prentice-Hall, Inc., Englewood Cliffs, New Jersey 07632

Library of Congress Cataloging in Publication Data

Craig, James V.
    Domestic animal behavior.

    Includes bibliographies and index.
    1. Domestic animals—Behavior. I. Title.
SF756.7.C73   636'.001'9   80-39803
ISBN 0-13-218339-0

Printed in the United States of America

10 9 8 7 6 5 4 3 2 1

Editorial/production supervision
    and interior design by Leslie I. Nadell
Cover design by Edsal Enterprises
Manufacturing buyer: John Hall

Prentice-Hall International, Inc., *London*
Prentice-Hall of Australia Pty. Limited, *Sydney*
Prentice-Hall of Canada, Ltd., *Toronto*
Prentice-Hall of India Private Limited, *New Delhi*
Prentice-Hall of Japan, Inc., *Tokyo*
Prentice-Hall of Southeast Asia Pte. Ltd., *Singapore*
Whitehall Books Limited, *Wellington, New Zealand*

TO MY WIFE, JEAN

*Her understanding, good humor,
and encouragement made this book possible*

Contents

Preface

This book is for those with the desire or need to know how domestic animals behave, why they do what they do, and the practical implications of their behavior. The subject is approached from both theoretical and practical viewpoints in ways that should be useful for students of animal agriculture and veterinary science, livestock breeders and producers, for those with companion animals, and, indeed, for anyone with a lively curiosity about animal behavior. Because domestic animals are dealt with, I have emphasized how man's care, genetic selection, and the physical and social environments provided by man affect behavior and thereby many aspects of animal well-being and performance.

Much scientific information concerning animal behavior has become available only recently and it continues to increase rapidly. Nevertheless, there are a number of topics for which the information available is only fragmentary and unifying concepts must yet be developed. General statements have been made where the level of knowledge appears to justify that, but I have attempted to avoid oversimplifying generalities otherwise.

It is assumed that the reader knows something of the general biology of animals, but even those without much training in this area should be able to understand much of what is presented. In dealing with relatively well-established subject matter, only a few classic studies and a sampling of recent investigations are cited. More references are given where much new information is available or where less-developed concepts are encountered. Those with access to well-stocked libraries can supplement their knowledge considerably by reading the articles cited for themselves.

Important clues concerning the basic functions of particular behaviors seen in domestic animals can be obtained by observing their feral or wild relatives in natural environments. However, this is not always possible, and even when it is, it is only a beginning; domestic animal behavior also needs careful scrutiny in the environments commonly provided by man. In addition, experimental treatments applied under carefully controlled conditions also produce valuable information. I have used a wide-ranging approach, including the use of information from all these sources, whenever possible.

This book is organized by topics rather than by comprehensive discussions of the behavior of different species in separate chapters. Nevertheless, statements about each of the major species are included under topics where reliable information is available or when it would be helpful for a better understanding of basic concepts. Brief discussions of actual studies and their outcomes are used extensively. This approach allows the reader to share some of the excitement and joy of discovery of the investigator as he searches for new knowledge. It also provides a basis for judging just how solid the evidence is from which hypotheses or conclusions are drawn.

ACKNOWLEDGMENTS

I acknowledge here my deep gratitude to A. M. Guhl, now deceased, who was a close colleague and friend during my first 20 years on the faculty at Kansas State University. His pioneering studies on the social behavior of chickens aroused my interest and his encouragement and cooperation contributed greatly to my continuing enthusiasm for doing behavioral studies. I have also had the good fortune of being associated with A. W. Adams and D. D. Kratzer in behavioral research and with a number of excellent graduate students. The latter are recognized indirectly by citing their research contributions where those are relevant to the purpose of this book. E. O. Price of the University of California (Davis) and J. L. Albright of Purdue University read the manuscript and contributed many helpful suggestions. Of course, they are not responsible for the shortcomings of this book. J. J. Lynch of the CSIRO, Armidale, Australia, allowed me to read a manuscript which he had prepared, prior to its publication, which was very helpful. D. L. Davis and G. H. Kiracofe kindly guided me in locating much useful material dealing with hormonal influences on reproductive physiology and behavior. A number of investigators willingly provided photographs and figures; their contributions are acknowledged within the figure captions and by references to their published research where that is pertinent. Leslie Nadell of Prentice-Hall was helpful in making needed editorial changes. To all these people I express my appreciation.

James V. Craig
Manhattan, Kansas

1

Introduction

Until agriculture was invented some 10,000 years ago, our ancestors lived as hunters and gatherers. In primitive societies knowledge of animal behavior is often necessary to sustain life, and people must cooperate when hunting large and dangerous animals. Our relationships with animals continue at the present; they entertain us and serve as companions, they work and provide transportation in many countries, they supply some of our clothing, and they provide the highest-quality foods available. For our own benefit an attempt should be made to understand them and their ways. In the process a better understanding of human behavior should result.

At about the time that crop cultivation became part of our way of life, domestication of animals also began (Zeuner, 1963). When considering animal behavior it is tempting to assume that those causes which influence human behavior are also responsible for animals' actions. The transfer of concepts between species may be useful if the species are closely related or under convergent selection, as will be discussed later. However, there is need for caution in interpreting animal behavior in terms of human experience. Seeing animals as if they are human can distract us and lead to faulty interpretations; this kind of approach is termed *anthropomorphic*.

Livestock producers in the past had practical knowledge of what their animals were likely to do in a given set of circumstances. They gained experience firsthand and learned, from those with special insight or previous knowledge, how to manage their stock under relatively natural conditions. Environments now provided for domestic animals are becoming more artificial, and the dam is being replaced in the care of young farm animals at earlier and earlier ages. Animals

1

are segregated by age, size, and sex, and human care is being replaced by mechanical devices. The need for reliable studies of behavior under these changing conditions is urgent.

What practical purposes are served by knowing more about animal behavior? Why do animals behave as they do? Why do individuals vary so little for some behaviors and so much for others? What genetic forces shape behavior and how important are they? How does early experience, age, sex, and the immediate environment influence what the animal does? Are there general principles that can help predict what will happen? In what follows some basic principles will be indicated and some of these questions will be answered. However, knowledge of animal behavior, and human behavior as well, is in a stage of early development, and some perplexing questions remain; it is both fascinating and frustrating to be on the cutting edge of the search into the unknown.

In writing this book I have taken the position that natural selection favors behaviors that increase chances for survival and reproduction and that information from wild and feral relatives provides clues about the origin of particular behaviors of domesticated animals. It will become apparent that some behaviors that are useful in natural environments may be maladaptive to the individual or the group in artificial environments.

To say that a particular behavior favors survival and eventual reproduction does not explain immediate causes. It is often valuable to know why certain behaviors occur. Sometimes a natural stimulus that brings out a desired activity can be mimicked by man. Occasionally, behavior can also be changed by modifying the internal state of the animal, as when hormones are fed or injected.

Our human population is producing profound changes in the environment. Many of those are recognized as unfavorable to the quality of both human and animal life. Husbandry that reduces labor and housing costs often results in physical and social conditions that increase behavioral problems. Some stressors acting on animals kept as sources of food may be unavoidable. Nevertheless, means of reducing behavioral stress are needed so that decreased labor and housing costs are not offset by losses in productivity. Also, for humanitarian reasons, there is a need to decrease the burden of the animal in serving man.

DETERMINANTS OF BEHAVIOR

What is seen when an animal is observed closely? Would it differ if the animal were living in a natural environment as compared to being in an enclosure or controlled by man? Ethologists, those who study animal behavior, especially in natural environments, often attempt to describe "species-typical" behavior. Sometimes they succeed to a large degree, but one of the most striking things about animals is the variation that exists between individuals; higher animals, in particular, show much variability when experience or learning enters the picture.

Figure 1-1 shows some of the major determinants of individual behavior. *Phenotype* is that which is seen, the particular behavior observed. Behavior is influenced by the animal's own set of genes (*genotype*), which is a unique combination possessed by no other animal unless it has an identical twin or is a member of a highly inbred line. Nevertheless, an animal has numerous genes in common with other individuals of the population of which it is a part, and its behavior is more like that of other members of its own population than it is to those of another group.

The gene pool of a population evolves under natural selection to provide the basis for generally adaptive behavior of its members under "natural" conditions. Because domestic animals are being considered, it is prudent to be aware that selection of any trait considered desirable may also influence behavior. Sometimes behavioral traits are selected directly. Effectiveness of selection, both natural and artificial, depends in part on preexisting genetic variation provided by mutations. In some cases, animals migrate from one population to another, thereby introducing a sample of a different gene pool. On the other hand, relatively small populations, which are isolated, are likely to have significant changes in gene frequencies because of random genetic drift.

Besides the genotype, the animal's physiological status, general environment, recent events, and presently occurring stimuli (or lack of such) also influence its behavior. Level of nutrition, seasonal effects such as day length and temperature, presence or absence of good health, earlier experience, and learning can all influence the activity seen.

THE ETHOLOGICAL APPROACH

Ethology, by definition, is the scientific study of animal behavior. Behavior, in turn, may be defined as the movements animals make, including change from motion to nonmotion, in response to external or internal stimuli (after Tinbergen, 1965). Ethologists have traditionally begun their studies by observing and recording details of a species' behavior in its natural habitat. Such a catalog of behavioral

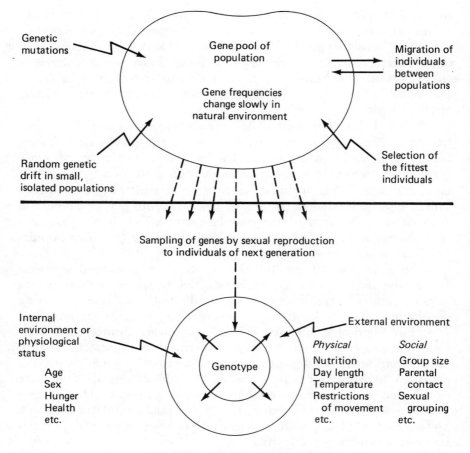

Phenotype: behavior of individual

Figure 1-1 Genetic and environmental determinants of a population and of the behavioral phenotype of an individual. Genetic
changes in the population's gene pool can be characterized in
terms of gene-frequency changes under the pressures of mutation, migration, random genetic drift, and selection. An individual's genotype is determined by chance, under random mating,
within the limits imposed by the population's gene pool. The
behavior of the individual depends on the interaction of its genotype with the external environment and on its physiological
status or "internal environment."

patterns is known as an *ethogram*. Insight gained from intensive observation often indicates the probable adaptive value of particular behavior patterns.

In addition to determining the function of particular activities of animals, ethologists gain insight into probable evolutionary processes by comparing related species and noting how differences in their environmental "niches" may have acted selectively in changing their genetic makeup influencing behavior.

Studies of seagull behavior involving parental removal of broken egg shells from the nest site soon after the young hatch and dry may serve to illustrate the ethological approach. Having observed such removal of shells by black-headed gulls, Tinbergen et al. (1962) supposed that this behavior functioned to reduce predation. The white inner surface of gull eggs reflects light efficiently and could serve as a conspicuous signal to crows and herring gulls, which prey extensively on young gulls and unhatched eggs, that food is nearby. This idea was tested by placing broken shells near nests containing eggs in certain areas and by being sure that none were present in other areas. A careful record was then kept of how many eggs were found by predators. As predicted, more were found when broken shells signaled their presence.

Tinbergen et al.'s study suggests how natural selection operates and how behavior changes (evolves) under such pressure. It seems likely that egg shell removal has not always been so important a part of the seagull's behavior. Predatory birds are presumably under continuing natural selection for keen eyesight. As soon as a parent bird began removing shells from the nest area, for whatever reason, the chance that the young would be preyed upon decreased. Removal of shells would, therefore, be a behavioral trait favored by nature. Parents acting in this way would have greater fitness; they would leave more offspring, on the average, than would parents lacking this behavior. Genes favoring the behavior of shell removal would tend to be passed on from generation to generation and would increase in frequency as compared to genes having no such effect.

REFERENCES

TINBERGEN, N., 1965. Animal Behavior. Time-Life Books, New York.

——, G. J. BROEKHUYSEN, F. FEEKES, J. C. W. HOUGHTON, H. KRUUK, and E. SZULC, 1962. Egg shell removal by the black-headed gull, *Larus ridibundus* L.: A behaviour component of camouflage. Behaviour 19:74–118.

ZEUNER, F. E., 1963. A History of Domesticated Animals. Harper & Row, Publishers, New York.

2
Natural Selection

FITNESS

Understanding the basic causes of animal behavior began with Charles Darwin and Alfred Wallace in 1858. They offered a theory of how nature could adapt a species to a changing environment by selecting, from a variable population, those individuals that were most "fit." An essential part of the theory is that traits determining fitness should be transmissible, at least in part, by biological inheritance. *Fitness*, as used here, refers to the relative ability to leave progeny in the next generation; it is a measure of reproductive success.

Natural selection deals primarily with changes occurring within species, but also, in a broader context, with survival of better-adapted species where species compete. Each species tends to fit a unique environmental niche because of a variety of physical and behavioral adaptations. That natural selection, acting within a species, does not always lead to its survival is evident from the vast number of species that have become extinct. It may be assumed, as a general principle, that those species which are most highly adapted to a specialized environment are most susceptible to extinction when the environment changes.

That significant evolutionary changes can occur over a moderate number of generations under favorable conditions has been widely recognized only within the past 20 to 30 years. It is particularly evident in insects. Kettlewell (1955, 1959) presented evidence of such changes from his study of moth coloration and behavior and the behavior of birds that prey upon the affected populations. Hundreds of species of moths that were predominantly light-colored a century ago have changed genetically so that they are now predominantly dark. Those changes occurred in areas where industrialization caused the deposit of a layer of soot on tree trunks and branches where moths sit motionless during the day. Camouflage coloration of moths provides protection from birds, which would otherwise feed upon them. Given a choice of background resting sites, moths choose those which match their own coloration. Kettlewell suggested that one segment of a moth's eye senses the background coloration while another segment senses its own color, allowing a comparison. When contrasts of sufficient strength are present, the moth is motivated to move, seeking a less contrasting background. Birds of various species were observed to prey selectively on those moths which contrasted with their resting site while missing others nearby which were protected by their cryptic coloration. In this example it is notable that even a century ago a low incidence of dark-colored forms was present (genetic variability was present) and that the environment was changed gradually as more and more of the countryside became blackened by smoke particles.

Lack of sufficient genetic variability or too sudden and drastic a change in the environment can lead to the total loss of populations. It is known that certain primitive tribes of man were wiped out by the arrival of more "civilized" colonists from other areas who brought along with them a heavy load of disease-causing organisms to which the migrants had developed some degree of genetic and passive resistance (the latter because of early exposure) but to which the natives were extremely susceptible.

It is important to note that behavior considered typical of a population may be changed not only by genetic selection but also by learning and social tradition. Although not evolutionary in the Darwinian sense, those changes may be described as due to *social inheritance*. Such changes may have occurred in coyote populations where animals survive in the presence of poisoned baits by living on killed prey only (F. R. Henderson, personal communication). Similarly, changes have been observed in the Japanese macaque monkey, in which the washing of foods has been found to spread as a cultural tradition.

Necessities shape behavior in the natural environment (Wilson, 1975). Finding food, mates, shelter, and avoiding predators are basic to survival and reproductive fitness. A successful life-style usually involves a balanced behavioral repertoire. Extremes of behavior are not likely to be successful. For example, a very nervous, timid animal may be best adapted to avoid predators, but too easily frightened to allow adequate feeding activity, mating with the opposite sex, and providing adequate care of the young. Clearly, such an individual is unlikely to be very fit in terms of transmitting its genes to the following generation.

POPULATION SIZE

Explosive Population Growth: An Example

It all or nearly all females of a population produce progeny at the level of their biological potential and if most of those survive and reproduce in like manner over several generations, a *population explosion* occurs. Such events are uncommon, but they do occur, and the effects can be profound. Myers and Calaby (1977) describe the explosive growth of a rabbit population which occurred in Australia. Rabbits are not native to that continent. Although European domesticated rabbits were introduced repeatedly from 1788 onward, the great outburst of population growth apparently stemmed from the importation in 1859 of 12 wild rabbits which were liberated in the state of Victoria. Myers (1971) suggested that the European wild rabbit evolved in an environment where predation was the primary factor maintaining population stability and that in Australia it "possesses no inbuilt physiological or behavioural mechanism to control its numbers."

The reproductive capacity and potential for geographic spread of rabbits is awesome. During the breeding season small groups form about a socially dominant male and female pair. The group protects a territory and excess animals are forced to move out, sometimes for distances of several kilometers. Rabbits also move to new areas whenever food shortages occur, usually after the breeding season. Sexual maturity is reached after 3 to 4 months, gestation lasts 30 days, and the young venture forth from the nest after 3 weeks. Myers and Calaby (1977) indicate that the number of young produced annually by a female varies from about 11 in marginal areas to 25 or more where conditions are favorable. Thus, in New South Wales their numbers could increase 10- to 12-fold within a single year! Zero population growth requires from 50 to 85% mortality. At any rate, wild rabbits thrive in Australia and they spread with alarming rapidity over much of the continent, as shown in Figure 2-1. The spread was aided at times by rabbit hunters and trappers, who "seeded" new areas in advance of the naturally occurring wave of spread.

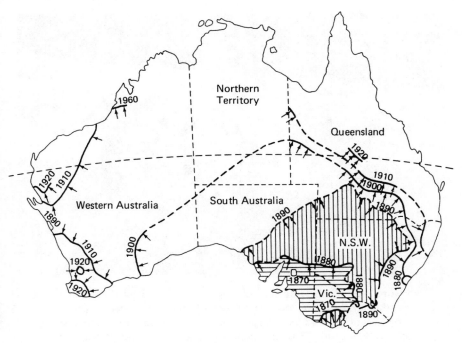

Figure 2-1 Pattern of spread of the rabbit of Australia. The release of a dozen wild European rabbits in 1859 resulted in an explosive expansion in numbers and geographic distribution. (From Myers, 1971; after Stodart, unpublished.)

The rabbit in Australia poses a serious economic threat to livestock producers by competing for vegetation and by degrading pastures. Some areas have been damaged to such an extent that the sheep-carrying capacity has been reduced to about half that of a century ago (see Figure 2-2).

Vigorous countermeasures designed to control the rabbit problem in Australia were relatively ineffective until after World War II. At that time the viral disease myxomatosis, highly lethal to the European rabbit, was introduced. Although the initial mortality rate was 99.9%, it had fallen to 40% by the mid-1970s in areas where annual outbreaks occurred. The highly virulent strains originally introduced are being replaced naturally by less virulent ones. (A disease organism does not benefit if it kills off all its host animals!) In addition, it appears that the rabbit population has developed some genetic resistance. Here we see a continuing example of how a natural selective force is in the process of altering a mammalian population in evolutionary terms.

Figure 2-2 Rabbit-infested country in southeastern Australia on the left side of the "rabbit-proof" fence has been largely stripped of ground-level vegetation. The field on the right has adequate pasture. The fence provides a flimsy and temporary barrier at best. (Courtesy of Australian News and Information Bureau.)

Regulation of Population Size

How are populations regulated in size so that relative stability is achieved? Limits imposed by food supply, disease, and predation have been recognized for at least a century. Recent studies indicate that social behaviors of many higher animals may also be important. Scott (1972:256) states that "Stable populations which have some kind of mechanism for controlling numbers can maintain themselves at high levels with optimum conditions for the existence of the individual members." This may be an overstatement, but it does reflect a change from the earlier view that life in natural environments is always harsh, and that starvation, disease, and predation are the only limiting factors of importance. The following behaviorally related phenomena are now recognized as deserving further study as regulators of population size.

Territorial Regulation. Territorial behavior involves the defense of a living area. Aggressive acts are usually directed against other members of the same species and sex which trespass. The male is far more likely to show this behavior than the female, although a mated pair or family group may act together to drive strangers away. Conspicuous displays and vocalizations commonly serve to keep other males from entering an occupied area without the necessity of physical encounters. Territoriality seems likely to increase the probability of successful reproduction, reduce predation, and ensure an adequate food supply for socially dominant males, their mate(s), and progeny. This kind of behavior is seen in many animals, from fishes through primates (Carpenter, 1958), but is not very common among the hoofed mammals, which account for many domesticated species.

Home ranges differ from territories in that areas occupied by individuals or social groups are not necessarily defended and may overlap. Nevertheless, when social conflicts occur, dominant animals behave as if they have a "portable territory" surrounding them. In similar fashion, subordinate animals may behave submissively or retreat as if they had overstepped a boundary when threatened or attacked by a dominant animal.

Territorial and home range behavior of feral fowl were observed on an uninhabited island off the coast of Queensland, Australia, by McBride et al. (1969). During the breeding season certain cocks (territorial males) were able to defend fixed boundaries and excluded adjacent territorial males. Some territorial males tolerated subordinate males, which were allowed to remain within territories but stayed at a distance and were generally excluded from mating activity. Young cockerels of very low status, (called "runts") usually left territories soon after dawn and returned at dusk; they moved inconspicuously along boundaries during the day. Dominant cocks were accompanied by harem flocks of 4 to 12 hens immediately prior to the breeding season (Figure 2-3). Although most males were not breeders, it appears that all hens of reproductive age were included within territories, as were hatched chicks sired by territorial males. It is not apparent that territorial behavior was involved in limiting population size of the feral fowl in this environment. Much mortality among baby chicks of second broods (there tended to be two broods per breeding season) occurred because of entanglement in sticky masses of berries. Predation was also an evident regulator of numbers, with most losses involving cockerels killed by feral cats or occasional human visitors to the island.

The role of territorial behavior in regulating population size in Red Jungle Fowl, believed to be the wild ancestor of domestic chickens, is unclear; adequate observation has been precluded to date

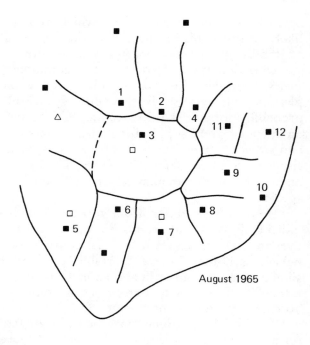

August 1965

Figure 2-3 Map of areas occupied by territorial cocks and their harems of hens at the beginning of the breeding season. Territorial males defended the boundaries shown and remained within their own areas. (After McBride et al., 1969.)

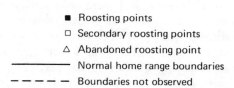

■ Roosting points
□ Secondary roosting points
△ Abandoned roosting point
──────── Normal home range boundaries
── ── ── Boundaries not observed

because of visual difficulties in the bamboo forests of southeast Asia, which is their natural habitat (Collias and Collias, 1967).

Precise information on the extent of territorial and home range behavior in wolves is generally lacking, but such behavior is believed to be of major importance in determining their fitness (Zimen, 1976). Such information is of considerable interest in understanding the behavior of dogs. Wolves are the most likely ancestors of dogs. Although dogs have been domesticated for more than 10,000 years, the basic behavior patterns of dogs appear to be more similar to those of wolves than those of any other canid (Mech, 1970:25–29). Eskimos have often used quarter-breed, half-breed, or even pure wolves for pulling sleds. Wolf–dog hybrids are fertile, as are wolf–coyote and dog–coyote hybrids, indicating their continuing genetic similarity.

Lone wolves, pairs, and small groups may live in and between pack-held territories by avoiding contact with the resident packs. Nonterritorial animals have reduced life expectancy and reproductive success. Lone wolves appear to live largely by feeding on leftovers of pack-killed animals and are often emaciated. Small packs have intermediate levels of food consumption and either fail to have pups or have no success in raising them (Jordan et al., 1967). Whether reduced fitness of nonterritorial wolves is caused primarily by inadequate food intake is not clear. Mech (1970) reported severe attacks on trespassing wolves by territorial pack members and also suggested that social stress could reduce reproduction in smaller packs. Territorial markings by urination and group howling could act as stressors on nonterritorial animals in the absence of direct physical contacts.

Low Social Status. Low social status effectively reduces male breeding success in feral or wild populations where males compete for mates in polygynous species (where one or a few males mate with several females each) as shown by McBride et al. (1969) in chickens, by Feist and McCullough (1976) in horses, and by Zimen (1976) in wolves. Such effects on male reproductive behavior may be of less importance in domestic animals under more controlled conditions, as will be seen in Chapter 17. Low-status males have more opportunity to mate when larger numbers of sexually mature females are present and when relative differences in status are reduced by having males of more uniform age and size present. It does not appear that differences in male social status is a very important regulator of population size under natural conditions, as individual high-status males are generally capable of impregnating a large number of receptive females.

Low social status of adult females is of considerable importance in controlling population size in wolves. Zimen (1976) obtained information by close observations of a captive group of wolves kept in a large enclosure (6 ha, about 15 acres) in a 5-year study. Only 16% of the assumed reproductive potential of sexually mature females was realized. No subdominant female had pups or was observed to copulate, although they seemed to exhibit sexual receptivity. Most low-ranking females were forced out of the pack during the breeding season, resulting in separation from male pack members. Even those subdominant females remaining in the pack failed to attract male sexual interest, which was concentrated on the dominant female. When a third-ranking male attempted to mate with a subdominant female in one year he was obstructed by the dominant female as well as by others. Zimen points out the general agreement of his results (*i.e.*, that only the dominant female of a pack usually has pups) with the results of other studies in wolf populations undisturbed by human interference.

Low female social status must be of less importance in regulating population size of species with lower reproductive potential than wolves. However, it is known that young heifers and ewes have shorter estrous or "heat" periods and are less aggressive sexually in seeking out males than are older and higher-ranking females. This aspect of behavior is considered further in Chapter 18.

Population Density. Crowding has been shown to be associated with physiological changes (Selye's stress syndrome) that reduce reproductive capacity in many species. This topic will be dealt with in some detail in later chapters but should be noted here as having important effects in regulating population size.

SECONDARY SEXUAL DIMORPHISM

Most mammalian and avian species can be easily classified into those that exhibit or lack marked secondary sexual dimorphism. Adult males of most ungulate (hoofed mammals) and gallinaceous (chickens, turkeys, etc.) species are much heavier than females (often 50% larger); have larger and more powerful forequarters; often possess weapons such as horns, tusks, or spurs; and may have prominent display characteristics, such as sickle tail feathers in cocks and fanning tail feathers in turkeys. In contrast, secondary sexual dimorphism is slight or even difficult to distinguish in many other species.

When ethograms of species with marked sexual dimorphism are compared to those without, some general differences are noted (Etkin, 1964). Those with marked dimorphism tend to live in moderate to large social groups with relatively few breeding males, breeding males possess harems (polygyny), and the young are often well developed at birth (precocial), depending primarily on maternal care. In contrast, species lacking dimorphism tend to live in nuclear family units with pair bonding of adult male and female, and the young are relatively slow in developing (altricial), requiring much parental care. Ducks and geese are exceptional; although pair bonds are formed between adult males and females and the sexes do not differ much in size, the young are well developed at hatching and soon follow their parents away from the nest.

The striking secondary sexual dimorphism in body weight of

polygynous as compared to lesser differences in pair bonding species is shown for several familiar animals in Table 2-1.

Why do such differences exist and what are the consequences? Apparently, males in sexually dimorphic species are large, powerful, and conspicuous because of the necessity to compete with other males for control of mating privileges. Biological success of an individual depends on transmitting genes to progeny. Males in polygynous, dimorphic species are successful in leaving more offspring if they are able to dominate other males, thereby assuring mating rights with a number of females for as much as a single season.

Table 2-1 *Actual and relative body weights of some familiar polygynous and pair-bonding species during adolescence or maturity*

Species and breed	Age	Weight (kg) Male	Female	$\frac{Male}{Female} \times 100$ (%)	References
		Polygynous species			
Cattle					Cole and Ronning
Holstein	Mature	1000	682	147	(1974:134)
Jersey	Mature	682	455	140	
Sheep					Cole and Ronning
Columbia	Mature	136	109	125	(1974:175)
Southdown	Mature	82	55	149	
Goat					Altman and Dittmer
Angora	4 years	55.4	31.7	175	(1964:98–99)
Toggenburg	4 years	70.9	51.6	137	
Turkey					Altman and Dittmer
Beltsville	30 weeks	9.1	5.1	178	(1964:101–102)
Bronze	40 weeks	14.8	8.4	176	
Eastern Wild	40 weeks	6.4	4.0	160	
Chicken					Altman and Dittmer
New Hampshire	20 weeks	3.4	2.3	148	(1964:101)
		Pair-bonding species			
Wolf					Mech (1970:12)
N.W. Territory (Canada)	Mature	44.5	38.6	116	
Alaska	Mature	38.6	32.3	120	
Ontario	Mature	27.7	24.5	113	
Goose					Altman and Dittmer
Graylag	16 weeks	5.4	4.7	115	(1964:101)
Duck					Cole and Ronning
Khaki-Campbell	Mature	2.0	2.0	100	(1974:187)
Pekin	Mature	3.5	3.0	117	

Such selection may lead to unphysiological specialization of males. Because the major function of males in polygynous species is to inseminate females and because they contribute little to care of the progeny, they are relatively expendible. If the dominant breeding male dies, he is easily replaced. In contrast, the fitness of females in polygynous species depends largely on their ability to provide adequate care of the young, thereby assuring survival of progeny. Under such circumstances males are likely to be inferior to females in life span (Daly and Wilson, 1978:70–76).

DIVERGENT AND CONVERGENT SELECTION

Two major types of change occur for species under natural selection. One is *divergent evolution* (including adaptive radiation), in which subgroups differentiate genetically from a common ancestral population in adapting to different ecological niches. The wolf and dog may serve as examples; the dog, separated as a subpopulation from wolves, has adapted to life as a companion of man (Scott and Fuller, 1965). Once begun, such differentiation may continue indefinitely. Thus, dogs continue to be subdivided by man's selection into different types, as shown in Figure 2-4.

The other type of long-term change is *convergent evolution,* in which animals separated in ancient times by divergent selection subsequently evolve in a similar direction under selection pressures favoring similar characteristics. Thus, both ungulates and gallinaceous birds survive and reproduce better under natural conditions by living in social groups rather than as solitary animals.

How have differences arisen between populations of various degrees of genetic disparity? Domesticated species are derived from mammals, birds, fish, and insects, the first two classes being more important. Obviously, large differences in anatomy, physiology, and behavior were favored by selection among the classes of animals which are fundamental to their survival in very different environments. Smaller differences exist between orders within classes (consider herbivores and carnivores within mammals) and between families within orders (horses, swine, and cattle within ungulates).

Species of different generic groups, within families, such as sheep and goats, show many similarities and yet are isolated as breeding populations. Certain species within the same genus or closely related genera may interbreed in some situations, providing viable, and at times vigorous, hybrids such as the mule (jackass

Figure 2-4 Dogs have become differentiated into distinct physical and behavioral types that predispose them to different roles in their relationships with man. A German Shepherd trained for police work is shown on the left and a Maltese, suitable as a pet, on the right. Maltese are only one-tenth the size of German Shepherds. (Photographs by the author.)

mated with a mare) and the "mule" duck (Muscovy drake mated with a duck of Mallard origin). Although such hybrids may be useful to man, they would probably not occur without our intervention and are essentially sterile.

Certain species classifications, although adopted by taxonomists, do not indicate an inability to hybridize. Thus, European and Indian cattle, classified as *Bos taurus* and *B. indicus,* respectively, cross readily and produce fertile progeny. Wolves, coyotes, and dogs, classified as *Canis lupus, C. latrans,* and *C. familiaris,* respectively, are also interfertile and produce fertile progeny.

INSTINCTIVE BEHAVIOR

The word "instinct" is sufficient to raise the ire of some students of animal behavior, particularly those who search for immediate causes of behavior under laboratory conditions. The word has been overused in the past as a means of avoiding explanations for causes of behavior which were unknown. Nevertheless, the word "instinct" embodies a useful concept and should not be abandoned simply because it has been misused previously. *Instinctive* (or *innate*) *behavior* is defined as behavior that increases fitness and occurs in adequate form when first needed. This kind of behavior is relatively stereotyped within species and is seen more in lower animals, which have little or no parental contact and short life spans. It is *not* implied that instinctive behavior is perfected when it first appears, particularly in higher animals, but that it is ordinarily adequate. The concept and importance of instinctive behavior and the possibility of its later modification, by the learning process, is considered in greater depth in later chapters (especially in Chapters 8 and 10).

Alcock (1979:51–62) and Dewsbury (1978:154–178) have carefully examined the positions initially taken by the classical European ethologists, who favored the widespread use of instinct for explaining most animal behaviors, and those of the American psychologists, who advocated discarding the concept. Those extreme views have been modified in recent years and it is now clear that each group has gained valuable insights from the other. As Dewsbury states: "While differences in emphasis remain, a synthetic study of animal behavior in which scientists of different backgrounds interact harmoniously appears to have come of age." He believes that there are, in fact, many species-characteristic behavioral patterns that may properly be termed instinctive, but that instinctive behavior should be viewed as being at one end of a continuum and learned behaviors at the other. Alcock also favors the view that many behavior patterns are a combination of innate and learned components.

Higher animals, although not so dependent on instinctive behavior as lower forms, also benefit from it. For example, the characteristic searching movements of newborn mammals are useful in locating the mammary glands of the dam, and the sucking and swallowing that follow are essential for initial food intake. Newborn and young animals of domesticated species also seek and maintain contact with warm objects and give distress calls if isolated, behaviors that generally improve their well-being and appear to be purely instinctive on their first occurrence.

On the other hand, numerous studies have demonstrated that many kinds of behavior, earlier termed instinctive, result from experience or may be greatly modified by experience. Thus, much maternal behavior in primates is absent or inappropriate if the mother was raised in isolation (Harlow and Harlow, 1962).

Instinctive behavior is not limited to the newborn in higher animals, although it is most prevalent at that time. It has been shown that inexperienced young animals of most species which have good vision and control of their bodily movements will avoid a "visual cliff." Experiments involve the presence of an apparent cliff, which is, however, not real, as a glass plate covers a precipice, as shown in Figure 2-5. As an example, young children who have control of their body movements will ordinarily refuse to move across such an area although anxious to rejoin a coaxing mother, even though they have presumably not experienced falling over an edge. Should danger of falling into a hole or over an edge arise, they would be protected. Young goats, representing a species adapted to mountainous habitats, show extreme avoidance of visual cliffs.

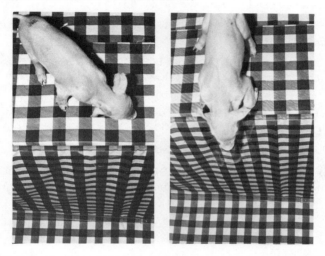

Figure 2-5 An inexperienced, 1-day-old pig instinctively avoids a sheer cliff. It is, in fact, protected from falling, as a sheet of glass covers the apparent dropoff. (Photographs by the author.)

REFERENCES

ALCOCK, J., 1979. Animal Behavior: An Evolutionary Approach, 2nd ed. Sinauer Associates, Inc., Publishers, Sunderland, Mass.

ALTMAN, P. L., and D. S. DITTMER, 1964. Biology Data Book. Federation of American Societies for Experimental Biology, Washington, D.C.

CARPENTER, C. R., 1958. Territoriality: a review of concepts and problems. *In* A. E. Roe, and G. G. Simpson (Eds.), Behavior and Evolution. Yale University Press, New Haven, Conn.

COLE, H. H., and M. RONNING (Eds.), 1974. Animal Agriculture. W. H. Freeman and Company, Publishers, San Francisco.

COLLIAS, N. E., and E. C. COLLIAS, 1967. A field study of the Red Jungle Fowl in north-central India. Condor 69: 360-386.

DALY, M., and M. WILSON, 1978. Sex, Evolution, and Behavior. Duxbury Press, North Scituate, Mass.

DEWSBURY, D. A., 1978. Comparative Animal Behavior. McGraw-Hill Book Company, New York.

ETKIN, W., 1964. Reproductive behaviors. *In* W. Etkin (Ed.), Social Behavior and Organization among Vertebrates. University of Chicago Press, Chicago.

FEIST, J. D., and D. R. McCULLOUGH, 1976. Behavior patterns and communication in feral horses. Z. Tierpsychol. 41: 337-371.

HARLOW, H. F., and M. K. HARLOW, 1962. Social deprivation in monkeys. Sci. Am. 207: 136-146.

JORDAN, P. A., P. C. SHELTON, and D. L. ALLEN, 1967. Numbers, turnover and social structure of the Isle Royale wolf population. Am. Zool. 7: 233-252

KETTLEWELL, H. B. D., 1955. Selection experiments on industrial melanism in the lepidoptera. Heredity 9: 323-342.

——, 1959. Darwin's missing evidence. Sci. Am. 200 (Mar.): 48-52.

McBRIDE, G., I. P. PARER, and F. FOENANDER, 1969. The social organization and behaviour of the feral domestic fowl. Anim. Behav. Monogr. 2: 127-181.

MECH, L. D., 1970. The Wolf: The Ecology and Behavior of an Endangered Species. Natural History Press, Garden City, N.Y.

MYERS, K., 1971. The rabbit in Australia. *In* P. J. den Boer, and G. R. Gradwell (Eds.), Dynamics of Populations. Proc. Adv. Study Inst. "Dynamics of Numbers in Populations," Oosterbeek, The Netherlands.

——, and J. H. CALABY, 1977. Rabbit. *In* Australian Encyclopedia. Grolier Society of Australia Pty. Ltd., Crows Nest, New South Wales.

SCOTT, J. P., 1972. Animal Behavior, 2nd ed. University of Chicago Press, Chicago.

——, and J. L. FULLER, 1965. Genetics and the Social Behavior of the Dog. University of Chicago Press, Chicago.

WILSON, E. O., 1975. Sociobiology. Harvard University Press, Cambridge, Mass.

ZIMEN, E., 1976. On the regulation of pack size in wolves. Z. Tierpsychol. 40: 300-341.

3

Domestication

Relatively speaking, man evolved only yesterday, and the domestication of animals is a very recent event in our history. Our knowledge of the past and of biological and physical processes, although obviously patchy, is growing at an astonishing rate. We are likely to think of the extinct dinosaurs as poorly adapted to life on this planet, but they flourished for over 150 million years, whereas recognizable ancestors of our own species came out of the dense forests and onto the open grasslands of Africa only 12 million years ago (Leakey and Lewin, 1977). By 2 or 3 million years ago our forefathers had advanced to the stage of making stone tools. Large game hunting became a part of our cultural heritage within the last million years; Reed (1974) estimates that event as occurring by about 750,000 years in the past.

Dogs were probably the first truly domesticated animals and are possibly the only species domesticated before crop cultivation was adopted. An early dog has been identified (Reed, 1974) from a site in Iraq that dates to 14,000 years ago. Sheep may have been the next animal domesticated, and they did not appear in that form until about 11,000 years before the present (Perkins, 1964). A crude calculation suggests that man has lived alongside domesticated animals for only about one-half of 1% as long as he has been a toolmaker.

Perhaps it is more than coincidence that large changes in Earth's climate and the recent "rapid" evolution of man occurred together. The major ice ages, alternating with warmer interglacial periods, have been with us for the last 2 million years or so. The last great ice age ended only 10,000 to 15,000 years ago, and as it ended, men became farmers and began the process of animal domestication.

Hale's (1969) definition of domestication is "that condition wherein the breeding, care, and feeding of animals are more or less controlled by man." Price and King (1968) view domestication as an "evolutionary phenomenon" involving changes in the gene pool of the population concerned. Clearly, the extent of such genetic changes required before a species is considered fully domesticated will vary. In terms of behavioral adaptations, some species are essentially preadapted to domestication (see below), whereas others are not. When the word *feral* is used, it refers to animals that were at one time domesticated but are no longer cared for by man. *Wild* animals are those that have never been domesticated. *Tamed* animals are those that have lost their fear of us. Animals in zoos are often in the transitional stage between being wild and domesticated, particularly if they have begun to reproduce in the captive condition.

The association of man and dog differed from the relationship of man and other species which were domesticated early. Wild dogs may have been attracted to man's campsites as scavengers, to feast on bones and scraps discarded by man, the hunter. Primitive peoples (even today) often adopt orphaned young animals, keeping them as pets. Thus, the adoption of puppies and similarities of life-style between man and dog, both adapted behaviorally by natural selection as group hunters, probably accounts for the symbiotic relationship that developed.

An early association was also formed with reindeer (Zeuner, 1963) which has been described as social parasitism (see Figure 3-1). Zeuner points out that human beings were required to be nomadic and probably exerted some control over the reindeers' movements and possibly over their reproduction. Tamed reindeer have been used as decoys in hunting untamed wild ones.

Although social parasitism may have occurred with herd animals other than reindeer, it seems less likely. Most authorities believe that true domestication of animals such as sheep, goats, and swine must have begun in a different way, after crop cultivation was initiated (Protsch and Berger, 1973; Reed, 1974). There is archaeological evidence that wild grains were harvested in Asia Minor for a period prior to systematic planting and cultivation (Zeuner, 1963). The human inhabitants of that area may also have created "wastelands" from forests, by burning, which may have served as pastures. Under such conditions, herbivorous mammals would tend to live close to man, providing opportunities for domestication.

Other centers of domestication included southeastern Europe,

Figure 3-1 Large herd of Lapland reindeer being "worked." Some Lapps still follow the old way of life—following the annual movement of the herds up to the mountains in spring and summer for calving, and back to their grazing grounds in the forest and along the coast in late autumn and winter. (Courtesy of The Swedish Institute.)

eastern and southern Asia, and the Americas. A rough approximation of time and place of earliest domestication is given for major species in Table 3-1. Estimates of the authorities cited differ for dates of first domestication of certain animals by as much as several thousand years. Nevertheless, the general pattern appears clear enough. The list of animals is incomplete, as it does not include a number of species of special importance in localized areas, such as the elephant, camel, yak, ass, and guinea fowl. (Those and others are discussed by Reed, 1974, and Zeuner, 1963.) Also absent are recently domesticated animals, such as the musk ox and laboratory animals, including the mouse, rat, gerbil, hamster, and various primates.

It is interesting to note how few of the thousands of animal species existing have been domesticated. Mammals and birds are the two classes primarily involved. Within mammals, the hoofed species are prominent, including

Odd-toed ungulates

Horse, ass or donkey

Table 3-1 *Apparent centers and dates of earliest domestication*

Species (probable order)	Geographic area(s)	Approximate years before present	References[a]
Mammals			
Dog	N.W. Europe, S.W. Asia, E. Asia	10,000–14,000	P&B, R, Z
Reindeer	N. Europe and Asia	15,000	Z
Goat	S.W. Asia	9000–11,000	P&B, R, Z
Sheep	S.W. Asia	6000–11,000	P&B, R, Z
Swine	Europe and Asia	5000–11,000	P&B, R, Z
Cattle	S.E. Europe, S. and S.W. Asia	6000–9000	P&B, R, Z
Water buffalo	S. Asia	5000	Z
Horse	Central Asia	4000–6000	P&B, R, Z
Cat	N. Africa	4000	P&B, R, Z
Llama and alpaca	N. South America	4000	Z
Rabbit	Europe	1000–2000	Z
Birds			
Chicken	S.E. Asia	4000	Z
Goose	S.E. Europe, N. Africa, E. Asia	4000	D, Z
Duck—Mallard	S.E. Europe and E. Asia	2000	D, Z
Turkey	S.W. North America	1000	Z
Duck—Muscovy	N. South America	1000	D, Z
Fish			
Carp	E. Asia	3000	B
Catfish	N. America	50	
Insects			
Silk moth	E. Asia	5000	Z
Honeybee	N. Africa	5000	Z

[a]Coded as follows: B, Brown (1977); D, Delacour (1964); P&B, Protsch and Berger (1973); R, Reed (1974); Z, Zeuner (1963).

Even-toed ungulates

Cattle, sheep, goat, Indian or water buffalo, musk ox, yak

Reindeer

Camels, alpaca, llama

Swine

Of poultry, there are two major groups, as follows:

Galliformes

Chicken, quail, pheasant, peafowl

Guinea fowl

Turkey

Anseriformes

Duck, Muscovy duck, goose, Canada goose

TRAITS FAVORING DOMESTICATION

Why are so few species domesticated? Certain behavioral character-istics are shared by most domesticated species; they are particularly evident in the ungulates and galliformes (gallinaceous birds). Nature has, in a sense, preadapted them to domestication. Of special impor-tance is the fact that these large, ground-dwelling birds and mammals exist in highly organized social groups. Stable dominance relationships are usually present, with the result that the animals are less likely to injure each other or to expend large amounts of energy in competing for resources that may be limited, such as food or mates. They can be kept together in relatively large numbers.

Promiscuous mating behavior is also characteristic of ungulates, and gallinaceous birds. Conspicuous secondary sexual dimorphism is common; adult males and females usually differ in size and strength, and in some species, in the possession or size of weapons such as horns, tusks, and spurs, and in display characteristics (reviewed in Chapter 2). A general consequence is that males are socially dominant to females, thereby facilitating mating. Under natural conditions polygyny is common (one male mates with several or many females). Such behavior is beneficial, as it is desirable in most livestock enter-prises to keep relatively few adult males.

Another behavioral characteristic of most domesticated species involves bonding of the mother to her young, which occurs soon after birth. If orphaned animals (sometimes of another species) are placed with the new mother at this time, she will often accept them as her own. Such fostering may have been important in the initial domestication of new species and is of continuing importance for the care of orphans of the animal's own species.

Precocial development at birth or hatching as found in the wild relatives of our common farm animals and poultry is also desirable. Not only are the parent animals freed of extensive care of the young, but the human caretaker can substitute for the natural mother either at birth or shortly after. A characteristic of species with precocial development is that a critical period exists early in life during which social bonds are formed. The presence of human beings during that period can result in *imprinting*. Being imprinted, at least partially, on man as well as on its own species can be beneficial to both the animal and its human caretaker, in that the animal's anxiety is reduced in the presence of man.

Most domesticated animals have been developed from species that were either herbivorous or omnivorous. As the human population continues to increase, it may become ever more important that domesticated animals do not compete directly with us for food. However, some animals (especially ruminants) are adapted to use roughages and other products that man cannot use; they supplement our needs by converting those to high-quality food.

Finally, it should be noted that domesticated livestock is adaptable to relatively large changes in climate and husbandry. In recent times, some highly artificial environments, such as colony cages holding many hens, have been found inadequate, although all physiological requirements were apparently met. Psychological requirements

Table 3-2 *Traits favoring animal domestication*

Desirable traits	Reasons
Social structure	
Matriarchal groups with stable "peck orders"	Aggressive acts and energy expenditure is minimal in acquiring necessities
Sexual dimorphism, male dominance, polygyny	Individual males impregnate a large number of females
Rapid bonding of mother to young	Mother and young stay together in groups; adequate care of young
Precocial development	
Senses function and movement and learning occur soon after birth	Young can join larger social groups soon after birth— important for protection from predators
Species bonding (imprinting, primary socialization) occurs soon after birth	Protects young from following inappropriate animals; allows early bonding to man and lack of avoidance response
Ingestive behavior	
Herbivorous or omnivorous; ruminant digestive system is desirable	Allows use of plants and waste products; reduces competition for food between animal and man
Adaptable to change	Absence of highly specialized requirements allows use in a variety of environments

may deserve greater attention as the environments provided become more artificial.

A brief summary of those traits considered as preadapting animals to domestication and their implications relative to caring for and managing livestock is given in Table 3-2. The wild ancestors of a number of domesticated species, such as dogs, cats, ducks, and geese, lack some of those traits, and their descendants vary in the extent to which they have been modified.

FITNESS REDEFINED

The fitness of an animal under domestication is determined largely by whether it pleases its owner enough that it is allowed to survive to maturity and reproduce. Nevertheless, natural selection may continue to operate under domestication. Man often values extreme or unusual types that would not survive in natural environments.

If biologically unfit types are favored by man, then artificial and natural selection oppose each other. As an example, "creeper" chickens are short-legged dwarfs caused by a dominant gene at a single genetic locus. Creeper hens were once thought to be superior for brooding chicks because of their short legs. At least one breed (Scots Dumpies) was developed, based on this simple genetic mutation. However, it soon became apparent that a serious biological problem was involved, as hatchability was poor and about one-third of the progeny had longer legs. Studies revealed that the creeper gene was lethal to the embryo when present in homozygous condition (*i.e.,* when both genes at the locus were alike for the mutant type).

A similar situation developed in beef cattle in the 1940s and 1950s. In that case "snorter dwarfs" (so called because of breathing difficulties) began to appear among the progeny of several important blood lines in the United States (Figure 3-2). Snorter dwarfs are not only grotesque in appearance, but move about awkwardly and have great difficulty, or may fail entirely, in reproductive behavior and the ability to complete gestation and deliver living young. Although the experimental evidence is not so clear as for the creeper chicken, it appears that the dwarf gene (in this case a recessive) was associated, when present with its "normal" allele, with what was considered to be the ideal body type at the time that they began to appear in significant numbers. Because of its recessive nature (with reference to dwarfing) and the apparent selection of heterozygous individuals as breeders, the dwarf gene spread widely in some breeds of beef cattle before the problem was recognized.

Figure 3–2 "Snorter dwarf" that appeared in the Angus breed in the United States. This female had a calf by caesarian section and later died of bloat when 4 years old. Such dwarfs reach about half of normal adult size, if they survive to that stage. (Courtesy of Horst Leipold, Kansas State University.)

BEHAVIORAL CHANGES UNDER DOMESTICATION

Behaviors appropriate in the natural environment may need to be modified if the animal is to be most useful under domestic conditions. Some changes are brought about by genetic selection and others by careful manipulation of the environment. The relative importance of genetic and environmental influences in changing traits is not always clear, but a brief description of some of the major changes associated with domestication is instructive.

Loss of Pair Bonding

Ducks, geese, wolves, and foxes form pair bonds under natural conditions. However, domesticated ducks and geese show promiscuous sexual behavior that allows males to mate with groups of females. In dogs, descended from wolves, individuals may have more than one mate. Foxes, kept as fur producers, may form pair bonds if together for more than a few hours, but prompt separation following mating allows a male to be mated with other females subsequently.

Loss of Broodiness

Under natural conditions chicken and turkey hens will lay eggs, generally on successive days, until the nest is full. The hen then becomes broody and begins to incubate. Daily removal of eggs tends to extend the laying period. With the advent of artificial incubators for chicken and turkey eggs, poultry producers began to see broodiness as a nuisance. Broody hens become protective of their eggs (and of any other eggs in a nest) and will often attack the person gathering them. In addition, broody hens need to be placed in an environment that lacks nests and nest-building materials to "break up" this behavior; otherwise, they may continue in the nonlaying phase for a lengthy period. Selection against broodiness has been practiced by leading commercial chicken breeders for a number of generations, and broody behavior is now rare in many highly productive strains, even when eggs are allowed to accumulate in nests.

A few years ago the Food and Agriculture Organization of the United Nations attempted to improve egg production in Indian villages by supplying White Leghorn cockerels (a highly productive egg-laying breed) to replace cocks of the common village fowl. The plan was to up-grade the stock by successive crosses to White Leghorn males. The plan was doomed to ultimate failure because the White Leghorn had become adapted to existence in a protected environment and has largely lost broody behavior among hens. Without artificial incubators, protection from predators, and vaccines for endemic diseases, the imported stock was rapidly lost.

Reduced Flightiness and Aggressiveness

Although substantial evidence is not available, it appears that our domesticated animals have, in general, become more placid than their wild relatives. Exceptions include breeds specificially selected for high levels of activity. Game breeds of chickens, race horses, cattle used for bull fighting, and breeds of dogs valued for their racing, guarding, or fighting behavior are among the notable exceptions.

Under domestication, animals are generally protected from predators and adequate supplies of feed are usually provided. The necessity for acute alertness and easy arousal into flight is no longer advantageous and may even result in injury because of running into walls or fences. Highly excitable animals may trample each other when kept in confinement by attempting to escape from harmless, but strange, objects and sounds. With food supplies provided, less energy is required for locating and ingesting nutrients. Aggressive behavior at the feed trough is also maladaptive to productivity of the group, as the need to compete is inappropriate when each individual is provided with an adequate supply.

Extension of Breeding Season

Most animals are seasonal breeders under natural conditions. The young are typically produced as the period of most abundant food supply approaches, thereby increasing their probability of survival. Domestic animals kept on range or allowed to forage most of the year are likely to be kept on a seasonal basis of reproduction, even if they have the ability to reproduce at other times. On the other hand, there is an increasing tendency for reproduction on a year-round basis as confinement housing schemes come into wider use. Increased specialization in different phases of production and a steady market demand over the year for food animals or their milk or eggs favor such a change.

Artificial lighting, temperature control, more adequate feeding, hormonal stimulation, and genetic selection have all been used in varying degrees to increase the length or to control the timing of the reproductive period. Chickens, ducks, cattle, swine, and certain breeds of turkeys and sheep can now produce progeny almost on demand if properly managed. Some breeds (as in turkeys and sheep) which are highly productive but seasonal breeders are under selection for extended duration of the breeding season. It appears that such genetic changes are possible, but estimates of the effectiveness of selection are relatively low for this trait and significant progress may require several generations of selection.

HARSH NATURAL ENVIRONMENTS

As the human population continues to grow at an explosive rate, animal scientists are exploring possibilities of using harsh natural environments as habitats for domestic animals. Although most domesticated species are adaptable to a wide range of conditions, the arctic region and large areas of Africa are not suitable for them. However, indigenous species may do well in those regions under conditions that forbid the use of animals evolved and selected in other geographic areas. Thus, musk oxen, superbly adapted to the arctic tundra, but near extinction because of man's hunting, appear to show considerable promise as a domestic species in Alaskan studies. In a very different environment, eland (a large African antelope) can be productive under conditions of extreme heat, dryness, and exposure to "sleeping sickness" disease which are lethal for cattle.

Crawford and Crawford (1974) have examined alternative systems of management of wild and domestic animals for both semiarid and high-rainfall areas in African ecosystems. Two broad arguments are advanced for incorporating new African species into domestic livestock systems; they have species-specific adaptations to heat stress, and they make excellent use of plants which are important to the water economy of the region. Behavioral adaptions are important. For example, eland rest in the shade during the heat of the day and feed on water-rich herbage in the cool hours of the night. They also reproduce and can be managed effectively in captivity, indicating that they are largely preadapted behaviorally to domestication. Related species presumably share those characteristics.

REFERENCES

BROWN, E. E., 1977. World Fish Farming: Cultivation and Economics. AVI Publishing Company, Westport, Conn.

CRAWFORD, S. M., and M. A. CRAWFORD, 1974. An examination of systems of management of wild and domestic animals based on the African ecosystems. *In* H. H. Cole, and M. Ronning (Eds.), Animal Agriculture. W. H. Freeman and Company, Publishers, San Francisco.

DELACOUR, J., 1964. The Waterfowl of the World. Vol. 4. Country Life Limited, London.

HALE, E. B., 1969. Domestication and the evolution of behaviour. *In* E. S. E. Hafez (Ed.), The Behavior of Domestic Animals, 2nd ed. The Williams & Wilkins Company, Baltimore.

LEAKEY, R. E., and R. LEWIN, 1977. Origins. E. P. Dutton & Co., Inc., New York.

PERKINS, D., Jr., 1964. Prehistoric fauna from Shanidar, Iraq. Science 144: 1565-1566.

PRICE, E. O., and J. A. KING, 1968. Domestication and adaptation. *In* E. S. E. Hafez (Ed.), Adaptation of Domestic Animals. Lea & Febiger, Philadelphia.

PROTSCH, R., and R. BERGER, 1973. Earliest radiocarbon dates for domesticated animals. Science 179: 235-239.

REED, C. A., 1974. The beginnings of animal domestication. *In* H. H. Cole, and M. Ronning (Eds.), Animal Agriculture. W. H. Freeman and Company, Publishers, San Francisco.

ZEUNER, F. E., 1963. A History of Domesticated Animals. Harper & Row, Publishers, New York.

4

Social Behavior
Effects on Spacing

Observation of birds and mammals in unrestricted environments reveals that their locations are determined not only by resources such as food, water, and shelter but also by the location of other animals. Spacing relationships will be considered in terms of individuals and of groups.

SPACING BEHAVIOR
AMONG INDIVIDUALS

Many factors influence distances between individuals. Although specific activities may not appear at first glance to involve social behavior and spacing effects, that may be misleading. For example, while eating a limited amount of food, distances between animals have been found to differ, depending on their relative social status. Although threats and avoidances are not obvious, those may be occurring at a very subtle level, or severe interactions may have occurred at a previous time so that the subordinate animal has learned to stay away until the dominant one has had its fill and moved away to digest its meal. Physiological status related to reproduction and parent–young interactions can also have profound effects. Thus, a sexually receptive female will ordinarily be closer to an adult male than will an anestrous female. Similarly, a mammalian mother whose udder is distended with milk will welcome the young to suckle, but will become refractory to continued nursing attempts when the milk supply is gone.

Hediger (1950) summarized the results of others and also carried out his own studies of wild animals in captivity. He made a number of important contributions to the understanding of spacing relationships between animals and between animals and man. McBride (1971) has also concentrated on theoretical aspects of spacing relationships and has added to our knowledge by observational studies as well.

Aggression or threats of aggression are commonly used in excluding others from an animal's *personal field* or personal space. McBride et al. (1963) showed that such personal space did not extend equally in all directions but was greater directly in front of hens and that most movements were concerned with avoiding the personal fields of dominant hens. Attacks or threatening behavior are most often terminated by the flight of subordinates to the edge of a dominant animal's personal space or by submissive postures or "out-of-context" behavior, such as juvenile food begging or female sexual presentation behavior by subordinates (even among males). Figure 4-1 illustrates postures and movements of wolves expressing differing levels of self-assurance and submissiveness and suggests the presence of personal fields.

Although most highly social species such as our common farm animals allow subordinates to remain in the group by exhibiting

Figure 4-1 Expressive behavior in wolves is highly developed. This encounter between three males clearly indicates the rank order and suggests the effects of personal fields on spacing. (From Zimen, 1976.)

submissive behavior, individuals may at times be injured or expelled from groups. The inability to escape or lack of submissive behavior can lead to the death of individuals in game chickens and breeds of dogs selected for fighting ability and aggression. Personal space requirements are likely to be minimal in noncompetitive situations, as when herbivorous animals graze on lush pasture, but increase when resources are limited.

Social distance refers to the maximum distance that a group-living animal will tolerate before moving toward others. Thus, it is associated with an affiliative social force, in contrast to the dispersal force reflected by personal space. Small social distances are typically associated with bonding and are most evident between adult male and female during the mating season and between mother and young during early life. Some species have very strong social bonds and animals may be greatly disturbed by separation from their groups. Ewes, for example, are gregarious and attempt to maintain visual contact with others while grazing (Crofton, 1958).

McBride (1971) points out that *living space* for gregarious animals lies "between the personal fields of neighbors and the social distance. Basically, it is the observance of these two distances which give a group its characteristic spatial architecture. . . ." When animals are under extensive conditions, McBride visualizes the living space of an animal as a broad neutral area between personal and social spaces.

Members of a group typically respond in two very different ways toward others of the same species, depending on whether the others are members of the group or are "outsiders." Although personal distances are maintained within the group, the existence of social distances indicates cohesiveness. Most or all adult members of a group usually react with aggressive behavior toward strangers intruding into the group. McBride describes the situation as follows: "Thus the group is a unit of lowered aggressiveness; other conspecifics are normally attacked should they enter this sphere. In this way groups remain discrete and maintain exclusiveness of membership."

Animals often have spacing relations with members of other species, especially when they compete with members of the other species for scarce resources such as food or shelter or when the other species is a potential predator. Very often a predator is treated by the prey animal in the same way as it would react to an extremely dominant and abusive conspecific: the predator is watched closely when approaching and flight occurs if it comes too close. Hediger (1950) used "flight distances" to measure this kind of event numerically and indicated how the behavior of animals can be modified in some ways, especially by animal trainers and zoo keepers, by knowing about spacing behavior. Movement of livestock usually involves approach by a handler or sometimes by a trained dog to within the flight distance of animals. Most animals, when being moved, are very sensitive to the direction in which the dog or man is facing or moving, so that their direction of travel can be changed by relatively small changes in the direction of movement of the herder.

USE OF SPACE BY SOCIAL GROUPS

Under domestication natural social groupings are seldom permitted. Relatively few intact males are ordinarily kept, young are likely to be separated from their dams before weaning would occur naturally, and animals are likely to be sorted and reassembled frequently. However, if it is known how naturally occurring social groups behave and organize themselves in space, then valuable insight is provided for understanding how behavioral problems arise under less natural conditions. Special attention will be paid to social groups that occur with regularity in wild relatives of domestic animals or in feral populations. The social groupings to be considered consist of breeding adults, parents, and progeny during the period of progeny dependence and those occurring during the remainder of the year.

Feral and Wild Fowl

Chickens, turkeys, ducks, and geese will be considered. They represent major avian species which are well adapted to diverse natural environments and also do well when brought under domestication. Chickens and turkeys belong to the same taxonomic order and share many traits, such as relatively large body size, marked sexual dimorphism, and the mating of a single male with several females. Ducks and geese belong to the same family of waterfowl. They differ most noticeably from chickens and turkeys in having much smaller differences in body size between males and females and in forming pairs during the mating season.

Chickens. The Red Jungle Fowl of southeastern Asia is believed by many to be the most likely ancestor of the common domestic fowl. Because its natural habitat consists largely of thick bamboo jungle and because of the extreme wariness of these wild birds, they have been studied little in their natural environment. Collias et al. (1966) observed unconfined Red Jungle Fowl in the San Diego Zoo, where they have become habituated to man. Those birds showed extreme territorial fixation. Dominant cocks associated with one to several hens, whereas subordinate cocks remained at a distance. Attractive sources of food caused some cocks to move into adjacent territories, but they were easily driven off by resident dominant males. The situation resembles that reported by McBride et al. (1969) for a feral population on an island of the Great Barrier Reef of Australia, as discussed in Chapter 2.

In addition to the defensive territorial behavior of cocks that occurs just before and during the breeding season, two other kinds of behavior involving use of space were reported (McBride et al., 1969). Hens that were incubating eggs and those with young chicks became solitary animals occupying "home ranges" or living areas. Home ranges of different hens often overlapped, and dominance relationships were clearly demonstrated between pairs of broody hens. Thus, although two hens might live in the same general area, one would typically avoid the other when the two met.

During the nonbreeding season, dominant males remained in fixed areas but did not have exclusive control of them. Subordinate males moved from group to group across boundaries but remained on the outer edges of groups that they joined.

Dominant cocks are important in determining group movements and in guarding against intruders. McBride et al. (1969:143) describe this:

> When a group moved it was the male who gathered the females together before moving. The hens maintained contact with him while moving, and he controlled the movement when it crossed open ground. When disturbed he gave the alarm call and walked parallel to the predator or potential predator while the hens quietly hid. When the flock was disturbed, males were actually observed to drive females away, by rushing toward them with wings spread. While hens fed, males spent the majority of the time on guard in the tail-up, wing-down alert posture. When he did relax to feed from time to time, his feeding often attracted the females to him. He was generally much more cautious than the females. He protected them from other males, which he drove away.

Broody hens with chicks behave in much the same way as dominant males; they control movements of the group, draw attention to food, maintain vigilance for intruders, and are protective in the presence of potential danger.

Turkeys. Wild turkeys are native to the New World. Large populations were present in what is now Mexico, and more than 10 million were estimated as being present in pre-Columbian times in what is now the United States (Schorger, 1966). They successfully exploited a variety of habitats, the only general requirements being the presence of cover for nesting, trees for roosting, food (usually including nuts or acorns), water, and winters not more severe than those of states such as Massachusetts. Although commonly thought of as forest-dwelling birds, they existed in large numbers in the eastern parts of the plains states from South Dakota into Texas. Destruction of much of their natural habitat and heavy hunting pressure, sometimes for the purpose of protecting grain crops, drastically reduced their numbers and eliminated them from much of their original range.

Flocks of wild turkey toms (males) move about together before the breeding season but tend to disperse in late winter as they become sexually active. Adult toms establish breeding territories on relatively open "strutting grounds," which may be free of trees or under trees where underbrush is minimal. Aggressive interactions occur and defended areas of several to many acres are set up; distances between displaying males are irregular but may be of the order of ¼ to ¾ of a mile (Dalke and Spencer, 1946). Territorial males attract hens by "gobbling" and "pulmonary puffs" and by elaborate movements and tail fanning (Figure 4-2).

Circumstantial evidence indicates that year-old wild toms are so inhibited by the presence of older males that they do not develop sexually in what could otherwise be their first breeding season. A. S. Leopold indicated to Schorger (1966:243) that in the absence of older toms, 1-year-old males show clear signs of sexual development and probably breed.

Hens are free to move about during the breeding season, and after approaching a gobbler in full display, invite copulation by assuming a sexual crouch. Although fertility from a single mating remains relatively high for several weeks, most wild hens visit the gobbler and copulate on most days before incubation begins. Hens

Figure 4-2 Wild male turkey in full sexual display on his strutting ground. Sexually receptive hens visit a tom on most days before incubation begins; "harem" size is usually about five. (Courtesy of W. M. Schleidt, University of Maryland.)

may stay in the gobbler's area for only a few minutes or for most of the day on such occasions. Males possessing strutting grounds usually have a harem of about five females.

As the spring weather moderates, hens hatch their broods in nests not far distant from the strutting ground. The young brood is soon led about in the vicinity until well developed, and the hen is alert and protective. If surprised with young before they are able to fly, she may "freeze" in attempting to avoid attention or may give a warning note (causing the poults to scurry for cover) and then may threaten, attack, or feign injury and move away to attract the predator's attention away from the poults. After a few weeks the young can fly, and hen and poults begin roosting together in trees. Hens and their poults may join together in bands and move about in their home ranges. The maternal–young bond appears to persist until the following breeding season.

Adult male and female turkeys spend much of the nonbreeding season in separate flocks, males in bachelor groups and hens in groups with their young.

Ducks. All breeds of domesticated ducks, except the Muscovy, are descended from the wild Mallard. Most adult Mallards begin forming loose male–female pair bonds several months before the spring migration (Bellrose, 1976:236), but continue to intermingle with the larger group during migration until arriving at the breeding grounds. Pairs then disperse and establish home ranges with central territories or core areas that are defended. Spacing between breeding pairs is primarily enforced by the drake, pursuing intruding hens of pairs flying over the territory, with occasional violent outcomes (Weller, 1964:50). The pair bond weakens and usually breaks during incubation. The drake then departs and joins other males in a moulting area.

From the foregoing description it might appear that ducks are "faithful" in their pair-bonded state, but drakes behave promiscuously if the opportunity arises; they direct courtship displays toward, pursue, and rape other females if such females are unable to escape and are not adequately protected by their own mates (McKinney, 1975).

Highly synchronized hatching occurs in ducks, so that all ducklings within a brood normally emerge in a period of 3 to 8 hours (Bjärvall, 1967). They are brooded in the nest for an additional 12 to 24 hours (Boyd and Fabricius, 1965). Ducklings then follow their mother away from the nest in search of their first meal. The duck is very protective of her brood and may be aggressive toward other birds, including ducklings of other females, or they may give a wing-flapping display to distract potential predators while the ducklings skitter off to hide in the nearest vegetation (Beard, 1964). Broods may travel 1 mile or more per day before the young are able to fly. Mother–young bonds tend to break as the ducklings learn to fly at 7 to 8 weeks of age and the mother then departs for a moulting area.

The yearly cycle of seasonal activities is completed with the autumnal migration of flocks to their wintering grounds. It is known that some adult Mallards pair with the same partner in more than 1 year, but how this is accomplished is uncertain.

Geese. The Graylag goose, ancestor of all modern breeds except the Chinese goose, was apparently domesticated more than 4000 years ago, as indicated by Egyptian frescoes, which show that their shape had already been changed (Delacour, 1964:155).

Geese establish territories centered on the nest area during the breeding season. They are well known for their fidelity, as mates frequently form pair bonds lasting for life. The male is also a faithful companion throughout the incubation and rearing phases. Weller (1964:68–69) indicates that no social hierarchy is apparent within a brood but that family units may interact. Thus, the members of a dominant family may chase a subordinate family. The bond between both parents and their progeny lasts through both autumn and spring migration and until the following breeding season, at which time the male parent drives away the young. The young tend to return each year to the place where they hatched.

General. Table 4-1 summarizes some of the behaviors of social groups and their use of space during the year under natural conditions. Each species is unique in several ways. Why do ducks and geese form pair bonds whereas chickens and turkeys do not, and why do those of geese persist throughout the year? It is reasonable to speculate that the gander must assist the goose if the young are to be reared, but clearly this is not the case with ducks, as the drake deserts the duck midway through incubation. Perhaps the large size and unphysiological specialization seen in cocks and toms cannot be tolerated by nature in species such as ducks and geese, which must migrate considerable distances twice a year.

Wild and Feral Ungulates

Ethograms of the wild relatives and of feral populations of our major domesticated species of mammals have been relatively slow in coming, but a few excellent studies are available. Wild and feral

Table 4-1 *Social groups and their use of space in feral populations or wild relatives of poultry during the annual reproductive cycle*

Species	Prebreeding and breeding season	Incubation and rearing season	Nonbreeding season
Chicken	Territorial cock and harem. Subordinate males tolerated at periphery of territory or excluded during day. Territorial male does most mating.	Hen becomes solitary; leads brood in overlapping home range area; broods kept separated until "weaned." Fixed dominance relations exist among hens.	Dominant cock remains in fixed area; others move across previous territories. Subordinate males avoid dominant cocks.
Turkey	Aggressive toms establish territories on strutting ground. Hens of harem move freely into and out of male's area. Nonterritorial males excluded.	Hen becomes solitary; leads brood; may join with other hens and broods as young mature. Dam-young bonds persist until next breeding season.	Males come together in bachelor flocks which are separate from those of dams and their progeny.
Duck	Loose pair bonds begin to form before spring migration. Pairs migrate with larger group. Overlapping home ranges with central defended areas on breeding ground. Drake pursues intruders, attempts to mate with females of other pairs.	Drakes move to moulting area; join other males as incubation proceeds. Duck is solitary with brood; keeps separated until young can fly; then departs to moulting area.	Flocks migrate to temperate climatic area for winter.
Goose	Pair bonds are long-lasting, often for life. Territory of pair is defended on breeding ground.	Gander remains with goose during incubation, and entire group lives together until next breeding season.	Parents and young continue living together. Male drives young away at beginning of next breeding season.

populations of sheep and horses do well and are relatively easy to observe when living in open environments. Wild and feral swine also thrive but are difficult to observe, as they are typically found in forests or seek cover when living in swampy areas. The last survivor of wild cattle or "aurochs" became extinct in Poland in 1627 (Zeuner, 1963:203) and relatively few truly feral herds exist.

Sheep. Sheep can subsist on a wide variety of plant material, including hard, abrasive, dry forage such as that found on high plateaus and mountain slopes (Geist, 1971:9–11), where sheep evolved. Because of their ability to use food of such low quality, their hardiness to climatic extremes, and their social system (especially their gregarious behavior) they are eminently successful as domesticated animals on a worldwide basis, being present in larger numbers than any other domesticated species of ungulates (Pope and Terrill, 1974).

There are 36 to 40 races of wild sheep, but considerable disagreement exists as to species classifications (Geist, 1971:6), and some researchers would lump them all together as a single species. Justification for that approach is based on the ability of all races tested so far to hybridize and produce fertile progeny. Nevertheless, it is more common to subdivide them into several species and to acknowledge descent of domestic sheep from *Ovis ammon,* the wild sheep of the Old World (Hulet et al., 1975).

Wild and feral sheep are typically organized as groups consisting of rams only and of dams and their juvenile progeny. However, during the breeding season rams disperse in irregular ways to join with sexually receptive ewes. Sheep become strongly attached to localities as well as to their social groups, and although migration between two or more home ranges may occur from year to year, particularly where seasonal changes are large, individuals tend to return regularly to fairly restricted areas. Particularly informative are the observations of Grubb and Jewell (1966) of feral Soay sheep on the Outer Hebrides island of Hirta (off the Scottish coast) and of Geist (1971) of mountain sheep in the North American Rocky Mountains.

The onset of the autumnal breeding season causes ram groups to move about and disperse in what appear to be strange ways. Some feral rams on Hirta departed from the intensively observed area, and others from fairly remote parts of the island moved in. During the height of the mating season many strange rams appeared, and some remained for up to 2 weeks before departing. The apparently strange migratory habits of wild rams was commented on by Geist (1971:88-89) as follows:

> The home range patterns of rams and the timing and distance of movements between their home ranges run as a rule so contrary to efficiency that had a human engineer designed them he would have been fired. . . . The long movements by rams away from, past, or through female groups to habitual rutting grounds, and their returns through perfectly good wintering areas to their accustomed winter home ranges, are not the exception but the rule.

He suggested that a young ram must develop attachment to specific home ranges from older rams that are followed during locality fixation of the young male. Young rams frequently switch from following one older ram to another, perhaps in this way establishing what later appears to be the irrational pattern of movements.

Grubb and Jewell did not include observation on agonistic and mating behavior in their report of the feral Soay sheep, but Geist presents extensive observations of those activities in mountain sheep. Rams spend considerable energy in violent head-to-head combat when strangers meet during "rut" or the breeding season and also when rams congregate in the spring. If strange rams differ considerably in horn size, which is generally indicative of age and strength, the smaller-horned ram may passively adopt subordinate status and submissive behavior. However, if the two do not differ greatly, they may clash in the manner shown in Figure 4-3. Mills (1937) observed rutting Bighorn rams with bloody noses, splintered horn tips, and injured legs. He reported that one ram died of injuries from fighting. Geist observed that other rams occasionally attacked from the side during a fight and wondered about the seriousness of internal injuries.

Older, more dominant rams of the wild mountain sheep do most of the mating and are preferred by females (Geist, 1971:174, 214-216). Nevertheless, some breeding is done by younger, less dominant rams if they can manage to chase a female away from the dominant ram and copulate before he catches up.

At the end of the breeding season each of the rams departs for his traditional wintering home range and all-male groups for the

Figure 4-3 Wild rams clash in head-to-head combat during the mating season. Older, more dominant rams do most of the mating. (Source: Geist, V., 1971. *Mountain Sheep: A Study in Behavior and Evolution.* Copyright © by the University of Chicago. All rights reserved.)

remainder of the year. Grubb and Jewell found that ram-group home ranges could cut across those of ewe groups. Ram groups are typically smaller than those of ewes. Older juvenile rams develop more and more independence from their dams and from the ewe group as a whole; they then begin following older rams.

Feral groups of ewes, averaging about 30 animals per group, showed fidelity to small home ranges throughout their lifetime on the island of Hirta. Grubb and Jewell thought that three diverse factors probably contributed: (1) natural features of topography and man-made structures (not in use) reduced freedom of movement [thus, a dry burn (creek) appeared to serve as a boundary] ; (2) particular areas of favorable grazing seemed to form focal points of "monopolized" areas; and (3) matriarchal families associated closely. It was thought that the attachment of a ewe lamb to her dam might last for a long time, possibly until the death of the dam. The results of a study by Hunter and Milner (1963) on pedigreed domestic South Country Cheviot sheep on a large hill pasture support the validity of the third factor. Older Cheviot ewes and their grown daughters were found to be closely associated during grazing and to restrict their movements to particular areas of a 102-ha (251-acre) pasture.

The concept of home range as related to sheep behavior has been elaborated on by Jewell (1966) and Grubb and Jewell (1966). Jewell points out that representing an animal's home range in the form of a map can be misleading, as certain areas may be heavily used and others only infrequently. Grubb and Jewell showed that when approaching or departing from a grazing area, sheep travel single file along well-defined tracks. Leaders of ewe groups are typically older animals.

Among the moutain sheep in Geist's study, rams could occupy as many as six to seven and ewes up to four home ranges. Only a minority of ewes occupied as few as two seasonal home ranges, the classical winter and summer ranges. Geist found that home ranges were anywhere from ½ mile to 20 miles apart and usually separated by one or more timbered gorges. Winter home ranges were usually less than ½ mile across, but in other seasons could occupy a large area of a mountain block. Mountain sheep return to the same home range year after year. Ewe groups were found to be far more constant in membership than ram groups.

The feral Soay sheep of Hirta had no predators to contend with, but Geist's mountain sheep were in the presence of wolves, coyotes, and bears. Living in a group has obvious advantages in such an environment. All sheep in a group were easily warned of the possible presence of a predator by the suddenly alert posture of a disturbed animal, which would raise its head and hold it rigidly, looking in the direction of the predator.

Ewes typically departed from their band for a few days while giving birth and establishing a bond with their lamb. For this event they usually moved to a secluded spot in rugged cliff areas, where they not only avoided contact with other sheep but appeared to be safer from predation as well.

Cattle. Although cattle are of major agricultural significance, relatively few observations of truly feral cattle have been made. However, Schloeth (1961) conducted a classic study on the "half-wild" Camargue cattle of southern France. Camargue cattle mate mostly from June to August, so that calves are dropped in the following spring when chances of survival are best. Schloeth states that no Camargue calves are born from June to December. Therefore, there are three yearly periods, which may be characterized as the calving season, the mating or rutting season, and the remainder of the year.

Hafez and Bouissou (1975) point out that among members of the bovine family, animals typically live in herds, and the behavior of individuals is greatly influenced by the other members. The most common group is the matriarchal herd, based on an old female, her adult female offspring, and their young. This appears from Schloeth's descriptions to be an accurate representation of the herd structure of the half-wild cattle he observed. Unfortunately for an ethological study, a number of the Camargue bulls were castrated at 3 to 4 years of age, and it is not clear to what extent the animals were further disturbed by human interference.

Groups of steers and bulls were observed by Schloeth to form groups, but those groups were not constant; they sometimes changed from day to day, with certain elements joining together or breaking away. Lone bulls or steers were observed only infrequently, and it appears that bachelor herds were the common social unit for males outside of the breeding season. Even so, cow and calf groups were usually accompanied by some males (usually young ones).

With the approach of the mating season several sexually mature bulls would join each cow–calf herd. Before the height of the mating season, 2-year-old males would attempt to mate, thereby causing long periods during which heterosexual fighting could occur, as only the weaker cows would submit to advances by 2-year-old bulls. Thus, it appears that social dominance of bulls over cows is necessary if mating is to proceed smoothly.

During the mating season each cow's urine was sniffed by bulls and steers, presumably in attempting to detect pheromones signaling the onset of sexual receptivity. As cows came into proestrus, all males were attracted and would attempt to "guard" the female. However, only the highest-ranking bull was likely to be in close attendance. After sniffing the genitals, the dominant bull would stand alongside the cow approaching heat, usually in a head-to-tail orientation. Dominant bulls would frequently leave the cow temporarily to drive subordinate rivals away, especially just before copulation. Cows in proestrus or estrus would also behave aggressively by chasing other cows away and would then return quickly to the bull, if he was not in close pursuit. High-ranking bulls and cows in full estrus maintained close, although temporary, pairings. After mating, the dominant bull would often lose interest and the cow might then be guarded by a lower-ranking male, which however, was unlikely to be allowed to mate. After the dominant bull had rested or grazed, he would usually return to the receptive cow and the full repertoire of guarding, intention movements, mounting, and mating would be repeated. Usually three or four matings occurred for a cow during her estrous period.

Following the rutting season, bulls tended to be found in male groups (see above), and herds made up primarily of cows and their calves persisted. Neither herds nor individuals of the Camargue cattle were observed to defend territories. Certain paths were used heavily, as between pastures, watering sites, resting areas, and so on. Herds tended to graze as discrete groups and did not appear to compete for grazing areas or for water except during periods of scarcity. Calves continued to suckle from their dams until about 2 months before the next calf was born.

During the calving season, the Camargue cows departed from the group to give birth. The calf remains hidden and apart from the herd during the first 3 or 4 days of life while the mother rejoins the herd, departing for only short periods to allow the calf to nurse. After the calf is led to the herd by its mother, it joins with the other young in a subgroup. The young calves tend to sleep and play together but do not go far from the cows during play. The mothers appear to nurse their calves at particular times, and most cows nurse their calves at the same time. When traveling with the herd, the calves move together. It is evident that the young form social bonds not only with their dams but also with each other.

Swine. Forest-dwelling pigs, ranging from the "wild boar" of Europe to a more refined animal with larger litter size in eastern

Asia, were domesticated soon after crop cultivation began (Zeuner, 1963). Although two and even three distinct wild ancestral species have been named, all domesticated pigs readily interbreed and produce fertile progeny, indicating genetic similarity. It appears likely that domestication occurred in numerous locations from local wild forms.

Because of their ability to eat all sorts of concentrated plant and animal foods, pigs readily revert to the feral condition throughout the world where winters are not excessively cold and they have access to food, water, and shelter. Feral swine are numerous in the southeastern coastal and southern areas of the United States. McKnight (1964), on the basis of an extensive survey, estimated their number to be in excess of 1½ million. They became so abundant in Queensland, Australia, that they were declared to be "vermin" and bounties paid for their extermination led to the slaughter of about 30,000 a year during the late 1940s (Pullar, 1950).

Keeping track of feral and wild pigs is difficult, but animals have been trapped, fitted with radio transmitters, and then followed by telemetry (Kurz and Marchinton, 1972). Adult boars join matriarchal herds when sows become sexually receptive. Apart from the breeding season, boars move about together, although older ones often range as solitary animals. Stegeman (1938) reported that wild boars visited farms in the neighborhood of their natural habitat, where they would often drive off or kill domesticated boars and breed with the sows. He concluded that the strongest, most able boars sired the most pigs. Kurz and Marchinton observed feral boars fighting in breeding groups, and a subordinate boar attempting to breed a receptive sow was driven away.

The wild boar population studied by Stegeman (1938) in the southeastern United States appeared to have two seasons of breeding yearly. Mohr (1960) and Gundlach (1968) reported that in years of good food supply, wild sows in Europe could have two litters per year, so that the reproductive potential was high during such periods of plenty. Feral sows are believed to produce litters on a year-round basis in moderate climatic areas, as expected, because of their relatively close genetic relationship with domestic stocks, which reproduce throughout the year.

Feral and wild sows approaching parturition typically separate from other swine and build a nest or shallow pit in a secluded spot where the baby pigs are born and kept for at least a few days. The sow subsequently leads her pigs away and may rejoin a small band usually consisting of not more than 8 or 10 mature females, and their young or the sow may simply reunite with her own juvenile progeny of previous litters. Larger groups of hundreds have been reported in areas of dense pig populations when the animals are disturbed. However, it appears that such large groups are not typical.

Horses. The fossil record of horse evolution is particularly well studied, but the history of domestication of the horse is still somewhat clouded (Waring et al., 1975). It seems likely that two races were primarily involved, the tarpan of the Russian steppes east of the Caspian Sea and the Przewalski horse of the Mongolian steppes. Zeuner (1963) supports recognition of the tarpan as the ancestor of most modern breeds. Although there were forest-dwelling horses in Europe prior to domestication, the wild horse appears to have been shaped by evolution to fit the steppe environment, consisting of semiarid, nearly level, treeless plains.

Wild tarpan horses became extinct in 1851 in the Ukraine (Zeuner, 1963:303) and Przewalski horses exist primarily in zoos; their fate in the wild seems uncertain. Feral horses, on the other hand, thrive in a wide variety of semiarid environments in the western United States and southwestern Canada when protected or ignored by man, their only serious predator (McKnight, 1964). It is conjectured that several million were present about a century ago.

Several studies of feral horses have recently been completed, but special attention will be given to those of Feist and McCullough (1975, 1976). Tyler's (1972) excellent observations of New Forest ponies in England are particularly fascinating when compared with those of Feist and McCullough, because, as they state: "In Tyler's population, males were removed by man, and a population of mostly females resulted. She observed many kinds of behavior, including vocalizations, being shown by females which we observed only in males. In our population, dominant stallions were extremely dominant, and apparently they suppressed a number of behavior expressions in females." The Pryor Mountain Wild Horse Range involved in the Feist and McCullough study, located in the western United States, included an area of about 13,600 ha (33,600 acres) of semiarid, varied terrain with some steep slopes. Two hundred and seventy horses were present, with a nearly balanced sex ratio, and all age groups were represented in this undisturbed population. Observations were limited to a 6 month period beginning May 1.

The breeding population centered on dominant stallions with harems. One to three mares and their immature offspring were typical in a group; the average social unit included five animals. Immature males within a group were extremely submissive in the close presence of the dominant stallion. Harem groups were essentially closed societies, and Feist and McCullough believed that all mating involved dominant harem stallions. Animals not belonging to a group were rejected by either the dominant stallion or the mares. Although 84 fights between stallions were recorded, only 12 of those were initiated as a result of a male attempting to add a mare to the group.

Excess stallions lived apart from harem groups either singly or in small groups, with an average group size of 1.8. Bachelor groups were socially organized and the dominant individual herded the other stallions in the same manner as harem stallions herded mares.

Home range behavior was clearly exhibited by the groups of horses observed by Feist and McCullough (1976:339-340). Each home range included at least one watering site and a large grazing area. There was considerable overlap of home ranges, as shown in Figure 4-4. Their areas varied greatly in size, depending on food resources and closeness to water. Plotting of 21 harem home ranges indicated areas of 300 to 3200 ha (740 to 7900 acres).

Several groups used the same areas and spacing between groups was controlled primarily by dominant stallions. They would approach each other with threats, which occasionally resulted in mild pushing and kicking matches. After such interactions, stallions would return to their harems and move them apart. An interesting result of encounters between groups was that during such encounters, animals within groups moved closer together. There did not appear to be defended areas or territories within the home ranges except for maintenance of linear spacing between groups. Horses approaching a watering hole would whinney, and if the site was already occupied by a band, would then wait their turn until the watering group moved off.

Foaling occurred during late spring and early summer in the Pryor Mountain horses. Although closely observed for foaling, no births were seen. Other studies (e.g., Rossdale and Short, 1967; Tyler, 1972) indicate that mares are able to control the onset of labor and that a high proportion foal at night. Feist and McCullough suggest that this may explain why mares were not seen giving birth.

Foals were able to follow their dams very soon after birth and stayed close to the side of the mother. Mares were obviously protective and threatened other horses approaching foals. Harem stallions also watched out for the welfare of foals and would herd them back to the group when they became separated. Mares without progeny were sometimes protective of another mare's foal.

General. As in the case of feral and wild ancestors of domesticated species of poultry, the species of hoofed mammals that have

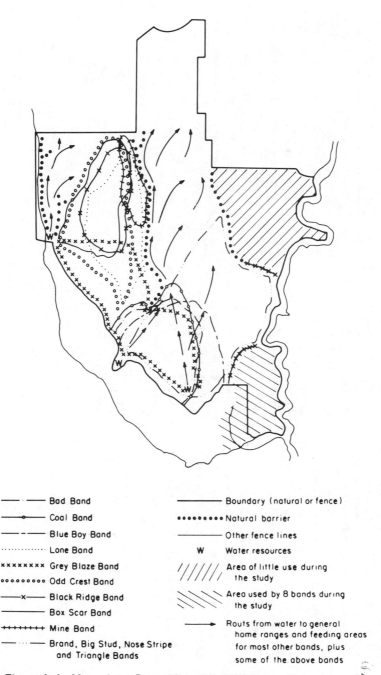

Figure 4-4 Map of the Pryor Mountain Wild Horse Range, showing home ranges of 21 harem bands. Note the large overlap of home ranges. (From Feist and McCullough, 1976.)

The legend of the figure reads:

— · — Bad Band
—○— Coal Band
— – – Blue Boy Band
············· Lone Band
×××××××× Grey Blaze Band
oooooooooo Odd Crest Band
—×— Black Ridge Band
——— Box Scar Band
+++++++ Mine Band
— ·· — Brand, Big Stud, Nose Stripe and Triangle Bands

——— Boundary (natural or fence)
•••••••• Natural barrier
——— Other fence lines
W   Water resources
////// Area of little use during the study
\\\\\\ Area used by 8 bands during the study
——→ Routs from water to general home ranges and feeding areas for most other bands, plus some of the above bands

Table 4-2 *Social groups and their use of space in feral and wild populations of ungulates related to domesticated species*

Species	Prebreeding and breeding season	Birth and rearing of young	Nonbreeding season
Sheep	Rams disperse from their male-only groupings to "traditional" breeding areas. Severe fights occur between rams of similar age and horn size for breeding privileges. Less mature rams do not fight with clearly superior rams, but may breed infrequently if ewes can be driven away.	Ewes separate from flock for a few days at parturition. Ewes and immature progeny form bands of relatives. Ram lambs become more independent of dam with advancing age. Ewe lambs tend to remain with mothers until death of dam.	Mature rams are found in bachelor flocks; several home ranges may be used. Adolescent and younger, mature rams follow older rams. Ewes and their progeny occupy overlapping home ranges with "monopolized zones" for matriarchal flocks.
Cattle	Several bulls are present, but breeding is largely restricted to socially dominant males. Subordinate bulls are sexually inhibited and interferred with if attempting to mate.	Cows separate from herd for calving; calves are kept hidden for a few days before being brought into the herd.	Bachelor groups form. (Relatively meager information is available.) Cows, their mature daughters, and immature progeny form matriarchal herds.
Swine	Boars join matriarchal herds when sows are sexually receptive. Reproductive period may occur once, twice, or throughout the year depending on climatic (and other?) conditions. Powerful, dominant males do most breeding.	Sow builds a nest and isolates herself with new litter. Litter remains in nest for a few days before following sow away. Sow and litter may rejoin immature progeny of previous litters, possibly including mature daughters and their progeny.	Boars are often solitary; range over wide areas. Sows and immature progeny form small matriarchal herds.
Horse	Dominant stallions and harems of mature mares plus their immature progeny form closely knit societies (average size, 5) occupying overlapping home ranges. Dominant stallions prevent mixing of groups by fighting if necessary. Excess mature stallions move about separately.	Mares foal mostly at night, rejoin herd promptly, but prevent contact of other horses with foal during early development. Stallion and mare of foal prevent foal from wandering away from the group.	Excess stallions move about singly or in bachelor herds.

been domesticated exhibit a variety of social behaviors regulating their use of space in natural environments. Nature is obviously opportunistic in selecting for whatever behaviors are adaptive in particular environmental niches. Some of the functions of particular group-related behaviors regulating the use of space are obscure; that is likely to remain the case where the ancestors of domesticated species have become extinct or where the environment has been drastically altered (usually by man) in recent times. A discussion of how wolf behavior is related to territory and home range is given in Chapter 2.

Table 4-2 summarizes some major aspects of the behavior of groups as related to the use of space by the major domesticated ungulates in natural environments.

REFERENCES

BEARD, E. B., 1964. Duck brood behavior at the Seney National Wildlife Refuge. J. Wildl. Manage. 28:492-521.

BELLROSE, F. C., 1976. Ducks, geese and swans of North America, 2nd ed. Stackpole Books, Harrisburg, Pa.

BJÄRVALL, A., 1967. The critical period and the interval between hatching and exodus in Mallard ducklings. Behaviour 28:141-148.

BOYD, H., and E. FABRICIUS, 1965. Observations on the incidence of following of visual and auditory stimuli in naive Mallard ducklings (*Anas platyrhynchos*). Behaviour 25:1-15.

COLLIAS, N. E., E. C. COLLIAS, D. HUNSAKER, and L. MINNING, 1966. Locality fixation, mobility and social organization within an unconfined population of Red Jungle Fowl. Anim. Behav. 14:550-559.

CROFTON, H. D., 1958. Nematode parasite populations in sheep on lowland farms. VI. Sheep behavior and nematode infections. Parasitology 48: 251-260.

DALKE, P. D., and D. L. SPENCER, 1946. The ecology and management of the wild turkey in Missouri. Mo. Conserv. Commun. Tech. Bull. 1.

DELACOUR, J., 1964. The Waterfowl of the World. Vol. 4. Country Life Limited, London.

FEIST, J. D., and D. R. McCULLOUGH, 1975. Reproduction in feral horses. J. Reprod. Fertil. Suppl. 23:13-18.

——, 1976. Behavior patterns and communication in feral horses. Z. Tierpsychol. 41:337-371.

GEIST, V., 1971. Mountain Sheep: A Study in Behavior and Evolution. University of Chicago Press, Chicago.

GRUBB, P., and P. A. JEWELL, 1966. Social grouping and home range in feral Soay sheep. Symp. Zool. Soc. Lond. 18:179-210.

GUNDLACH, H., 1968. Brutfürsorge, Brutpflege, Verholtensontogenese und Tagesperiodik beim Europäischen Wildschwein (*Sus scrofa* L.). Z. Tierpsychol. 25:955-995.

HAFEZ, E. S. E., and M. F. BOUISSOU, 1975. The behaviour of cattle. *In* E. S. E. Hafez (Ed.), The Behaviour of Domestic Animals, 3rd ed. The Williams & Wilkins Company, Baltimore.

HANSON, R. P., and L. KARSTAD, 1959. Feral swine in the southeastern United States. J. Wildl. Manage. 23:64-81.

HEDIGER, H., 1950. Wild Animals in Captivity. Butterworths Scientific Publications, London.

HULET, C. V., G. ALEXANDER, and E. S. E. HAFEZ, 1975. The behaviour of sheep. *In* E. S. E. Hafez (Ed.), The Behaviour of Domestic Animals, 3rd ed. The Williams & Wilkins Company, Baltimore.

HUNTER, R. F., and C. MILNER, 1963. The behaviour of individual, related and groups of South Country Cheviot hill sheep. Anim. Behav. 11:507-513.

JEWELL, P. A., 1966. The concept of home range in mammals. Symp. Zool. Soc. Lond. 18:85-109.

KURZ, J. C., and R. L. MARCHINTON, 1972. Radiotelemetry studies of feral hogs in South Carolina. J. Wildl. Manage. 36:1240-1248.

McBRIDE, G., 1971. Theories of animal spacing: the role of flight, fight and social distance. *In* A. H. Esser (Ed.), Behavior and Environment. Plenum Press, New York.

——, J. W. JAMES, and R. N. SHOFFNER, 1963. Social forces determining spacing and head orientation in domestic hens. Nature 194:102.

McBRIDE, G., I. P. PARER, and F. FOENANDER, 1969. The social organization and behaviour of the feral domestic fowl. Anim. Behav. Monogr. 2:127-181.

McKINNEY, F., 1975. The behaviour of ducks. *In* E. S. E. Hafez (Ed.), The Behaviour of Domestic Animals, 3rd ed. The Williams & Wilkins Company, Baltimore.

McKNIGHT, T., 1964. Feral livestock in Anglo-America. Univ. Calif. Publ. Geogr. 16:1-87.

MILLS, H. B., 1937. A preliminary study of the bighorn of Yellowstone National Park. J. Mammol. 18:205-212.

MOHR, E., 1960. Wilde Schweine. A. Ziemsen Verlag, Wittenberg, Lutherstadt.

POPE, A. L., and C. E. TERRILL, 1974. Sheep, goats, and other fiber-producing species. *In* H. H. Cole and M. Ronning (Eds.), Animal Agriculture. W. H. Freeman and Company, Publishers, San Francisco.

PULLAR, E. M., 1950. The wild (feral) pigs of Australia and their role in the spread of infectious diseases. Austr. Vet. J. 26:99–110.

ROSSDALE, P. D., and R. V. SHORT, 1967. The time of foaling of thoroughbred mares. J. Reprod. Fertil. 13:341–343.

SCHLOETH, R., 1961. Das Sozialleben des Camargue-rindes. Qualitative und quantitative Untersuchungen über die sozialen Beziehungen—insbesondere die soziale Rangordnung—des halbwilden französischen Kampfrindes. Z. Tierpsychol. 18:574–627.

SCHORGER, A. W., 1966. The Wild Turkey: Its History and Domestication. University of Oklahoma Press, Norman, Okla.

STEGEMAN, L. C., 1938. The European wild boar in the Cherokee National Forest, Tennessee. J. Mammol. 19:279–290.

TYLER, S. J., 1972. The behaviour and social organization of the New Forest ponies. Anim. Behav. Monogr. 5:87–196.

WARING, G. H., S. WIERZBOWSKI, and E. S. E. HAFEZ, 1975. The behaviour of horses. *In* E. S. E. Hafez (Ed.), The Behaviour of Domestic Animals, 3rd ed. The Williams & Wilkins Company, Baltimore.

WELLER, M. W., 1964. The reproductive cycle. *In* J. Delacour, The Waterfowl of the World. Country Life Limited, London.

ZEUNER, F. E., 1963. A History of Domesticated Animals. Harper & Row, Publishers, New York.

ZIMEN, E., 1976. On the regulation of pack size in wolves. Z. Tierpsychol. 40:300–341.

5

The Senses and Communication

Tinbergen (1965) points out that "different animals, including man, have different windows to the world" and "each perceives best only that part of the environment essential to its success." People rely heavily on sight, and we tend to assume that other animals do the same. The fossil record suggests that early man evolved in the semiarid, open grasslands of Africa, where, by standing erect on two legs, he could search his habitat visually for food sources and watch for dangerous predators. Predators have bifocal vision, as we do, which allows excellent form and depth perception and accurate estimates of distance to objects. Many birds and man have color vision, but most ungulates do not or it is only poorly developed. Grazing mammals and birds that feed primarily on plant foods have wide-set eyes and an essentially panoramic view of the world. They can see in virtually all directions at the same time, except for a small area directly behind the head. Panoramic vision is an adaptation for survival in prey animals; it decreases the probability of being surprised by a predator approaching from a blind spot. Nevertheless, such animals typically take advantage of their limited area of bifocal vision by turning their heads to look directly at disturbing objects after they are located to the side or rear.

Hearing is well developed in most domesticated animals. It is known, for example, that dogs can hear whistles pitched so high that they are inaudible to the human ear and that horses have an extremely keen sense of hearing.

The sense of smell is poorly developed in poultry and other birds, but most mammals have olfactory sensitivity far superior to our own. Dogs that follow scent trails are obviously well equipped and are generally believed to have at least a hundredfold more sensitive olfactory receptivity than man. Most mammals use the sense of smell for social recognition and for detecting sexual readiness. Territorial marking by urination, defecation, or rubbing of scent glands on conspicuous objects is well recognized in dogs, cats, and many ungulates. Most herbivores avoid plants that are detected by olfaction as having been contaminated by elimination, although sheep may seek out and consume "urine patches" because of their lush growth in otherwise closely grazed pastures.

Sensitivities to taste and tactile (touch) stimuli vary widely among species. Those will not be dealt with here, except to note that food preferences are obviously influenced by flavor, tenderness, and so on, and that most mammals are soothed either by self-grooming or by the grooming received from others.

COMMUNICATION

When information is exchanged among individuals, there is communication. In the human context communication is usually in terms of written or spoken words, but it can go far beyond that. Words need not be used in communicating anger, happiness, sorrow, or fear. Information is also received by the sense of touch when either physical aggression occurs or carresses are delivered. Because of human dependence on words, we are likely to be less sensitive to other means of sending and receiving information. However, the loss of sight or hearing clearly forces intensified use of the other senses. Animals, although not having the use of words, appear in many cases to have highly developed alternative routes of communications. People who work closely with them are often able to judge their state of arousal and to anticipate their behavior by being attentive to the signals being transmitted.

Schein and Hale (1965) studied stimuli associated with sexual behavior in animals and classified those into three categories. The kinds of stimuli cited by them represent information transmitted and received. Their system was applied specifically to sexual behavior. One additional category (mood and intention signals) will be added and the system expanded to include communications relating to other social behaviors as well.

Broadcasting

Information is broadcast when an adult turkey tom "gobbles" on his strutting ground or when sexually receptive mammalian females excrete odorous pheromones in their urine (Figure 5-1). When a calf, lamb, or chick has become separated from its mother or familiar companions, it gives a distress call. Broadcasting occurs whenever an individual advertises its location by such signals and thereby invites contact with appropriate animals. Broadcasting is also evident when dominant or territorial individuals make themselves conspicuous (by any means) in order to intimidate others from approaching or competing for a scarce resource.

Identification

An approaching individual is typically identified before any other social interaction occurs. Means of identification vary depending on distance, species, sex, physiological status, and experience.

Hens that have an established peck order readily recognize all other individuals within small flocks. Because birds have a poorly developed sense of smell and the frequency of physical contacts are much reduced in organized flocks, they must be using visual or auditory clues. Guhl and Ortman (1953) suspected that visual clues were most important; they cited Schjelderup-Ebbe's observation that when the loppy comb of a hen was turned and fastened to the

Figure 5-1 A young bull shows the olfactory reflex or "flehmen" (lip curl) associated with inhalation through the upper respiratory passages. Although not restricted to mature male ungulates, it appears most commonly in them during the mating season and is usually exhibited when examining the urine or genitals of females in testing for pheromones secreted during proestrus or estrus. (Photograph by the author.)

other side of her head, she was treated as a stranger. Guhl and Ortman altered the appearance of hens by various means and then rein-
troduced them into their flocks (Figure 5-2). They concluded that
individual recognition among hens depended primarily on features
of the head and neck, as alterations in appearance of that area resulted in attacks by subordinate hens that failed to recognize their
dominant penmates after such alterations were made.

That hearing is a means of individual identification in precocial
birds was demonstrated by Ramsey (1951) with chicks, turkey
poults, ducklings, and baby pheasants; young birds of those species
were able to find their foster mothers, which were concealed in
boxes, even though other foster mothers were calling from other
boxes. That recognition was learned rather than instinctive was
indicated in Ramsey's study when young birds, only a few days old,
avoided strange adult females of their own species, but followed a
foster mother of a different species after bonding to the foster
mother had occurred.

Domesticated mammals typically use their sense of smell at close
range and visual and auditory characteristics when farther away for
identification of individuals. Those who have befriended a dog have
probably observed that as they approach the animal it may give some
sign of recognition, but usually continues to be vigilant. A spoken
word will usually set the tail to wagging. But final confirmation,
involving a sniff of hand or leg, causes the animal to relax fully, re-
assured that the person is, indeed, a familiar companion.

When a strange animal is encountered, it is important that its
sex be identified. Behavioral cues appear to be important in making
this discrimination. Domm and Davis (1948) made this very clear in
their study of sexual behavior in domestic chickens. When unac-
quainted birds of the same sex are placed together, they commonly
face each other, lower their heads, and ruffle their neck feathers. A
fight may then take place to establish dominance. When a cock is
placed with a strange hen, she will usually either assume a sexual
crouch when approached or will actively avoid him. Hens rarely
attack or threaten a strange, full-grown male as they would a strange
hen; presumably his much larger size and appearance (large comb
and wattles, etc.) readily establish his dominance and indicate sex
without need for combat. Young cockerels, capable of successful
copulation and fertilization, may be severely attacked by older,
larger hens when introduced into a flock (Grosse and Craig, 1960);
they are challenged and attacked as a strange hen would be.

Domm and Davis (1948) commented on the function of "waltz-

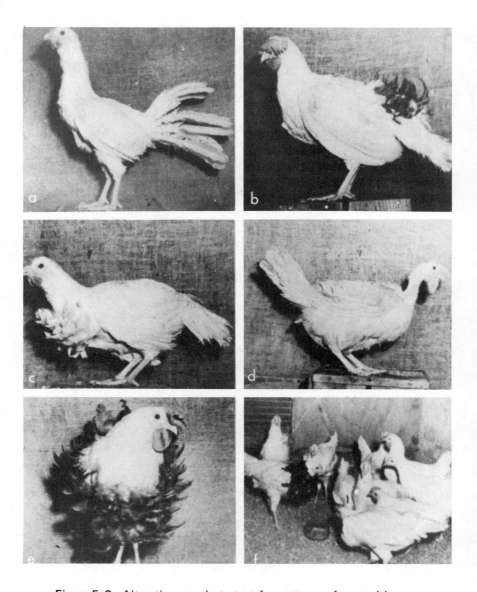

Figure 5-2 Alterations made to test for patterns of recognition: (a) change in contour by addition of white feathers to tail; (b) change in contour and color by addition of red feathers to saddle region; (c) contour altered by addition of white feathers to breast; (d) denudation of the neck to alter contour; (e) trunk modified by addition of red feathers to alter contour and color; (f) penmates avoided a pullet after her neck was disguised by addition of red feathers. (From Guhl and Ortman, 1953.)

ing" behavior, which has been described by some as courting behavior. They observed that it is likely to occur when a rooster encounters a strange chicken which fails to behave in the manner typical of another male (raised hackle feathers, aggressive posture) or of a receptive female (sexual crouch). In this circumstance the rooster dances around the other bird with the primary wing feathers spread, lowered, and sometimes scraping the floor and shank (Figure 5-3). If the other bird squats, he copulates; if it indicates aggressive intent, he fights with it. Thus, the waltz is an ambiguous activity, as it contains elements of both sexual and aggressive behavior and forces behavior of the other individual which is then used for identification of sex.

Figure 5-3 Waltzing behavior by a cock is directed toward a hen. This display is most frequently seen when a cock encounters a strange adult that fails to behave in a manner typical of either male or female. It contains elements of both sexual and aggressive behavior and usually forces a behavioral response of the other individual which is useful in indicating its sex. (Photographs by the author.)

It is relevant to note that males or castrated males of several species which are artificially restrained in the position of a sexually receptive female are typically mounted by sexually motivated males. Steers may be used to stimulate sexual mounting for the artificial collection of semen in bull studs.

Does submissive behavior by subordinate males in all-male groups lead to confusion as to sexual identity? Observations of Feist and McCullough (1975, 1976) could lead to such an interpretation. Dominant stallions in bachelor herds behaved in some ways as if they were dominant males with harems of females. Subordinate stallions were herded, kept separated from other groups, and had their fecal matter covered by the dominant stallion's own as if they were mares.

Observations on all-male flocks of mountain sheep also suggest that dominant males may be confused as to sexual identity of subordinates that exhibit behavior typical of females (Geist, 1971:130–132). Geist indicates that there is no typical female form in sheep and that male groups "were homosexual societies in which the dominant acts the role of the courting male and the subordinate the role of the estrous female." He states: "One can claim that the young male acts like an estrous female, mimicking her behavior and appearance, which allows him to live side by side with larger males." An alternative explanation given by Geist is that the female during estrus acts like a subordinate young male.

Early experience, particularly during the critical period for primary socialization or imprinting, can have profound effects on identification. Correct identification of an animal's own species depends on exposure during this period. Birds, in particular, if in close and exclusive contact with a foreign species during their primary socialization stage, may direct their sexual courtship during adult life toward an inappropriate species. Thus, male turkey poults imprinted to human beings, although tame with other turkeys, court people in preference to other turkeys. They identify both people and turkeys as appropriate sexual partners, but prefer people (Schein and Hale, 1965:443). Such incorrect associations appear to be more easily reversed in mammals, as discussed in Chapter 9. Nevertheless, Hediger (1950:168) indicates that tame males of the roe deer mistake man as a rival during the rutting season and attack people vigorously, as they would attack another male deer. A male moose, socialized to man, was reported by Hediger as making sexual advances as if the keeper were a female moose; such situations can be extremely dangerous.

Mood and Intention Signals

Most descriptions of behavioral signals indicating level of arousal, mood, and intention involve companion or working animals, such as dogs, horses, and cats, or those that can be studied in laboratory settings, such as chickens. Visual and auditory signals are commonly involved and are likely to occur at the same time, especially if the animal is highly aroused. Thus, a dog preparing to attack indicates its aggressive intent not only by a menacing growl but also by visual signals.

Body language or visual signals may involve movement of the entire body as well as its parts. Motion pictures, photographs, and sketches often serve well for descriptive purposes. Ambivalent signals can be conveyed when an animal has conflicting drives, as when a dog is behaving aggressively, even though it is frightened (Figure 5-4).

Signal–Response Sequences

In flocks and herds where adult males and females live freely together, mating behavior involves regular sequences of events between the sexes. A particular behavior by one sex either elicits a positive or a negative response by the other. When positive responses follow each other without interruption, a successful mating occurs. It appears that mating behavior patterns are largely innate responses to "social releasers" or stimuli provided by the partners. This interpretation is favored by the rather stereotyped nature of the responses and by the fact that animals raised in isolation are often able to mate successfully, although awkwardly (Harper, 1970). Experience improves the mating behavior of previously isolated males; their difficulties are usually related to faulty orientation, as they may initially attempt to mount the female from the side or front. Gilts reared in isolation by Signoret (1970) were attracted to boars when in estrus and did not differ in copulatory behavior when compared to normally reared females.

The first phase of mating involves courtship by the male. Roosters, turkey toms, and males of other gallinaceous species put on spectacular courtship displays (consider the peacock!), as judged by the human observer, although the female of the species may appear to be rather unimpressed. Ungulate males actively court females, primarily by tactile means. They may nudge, sniff, playfully bite about the head and neck, and lick the female's genitalia. They also encourage the female's attention by secretion of pheromones and by

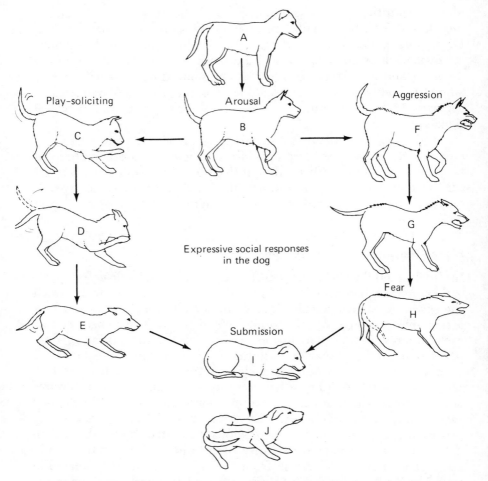

Figure 5-4 Schema of body postures and facial expression in the dog related to changes in motivational state. A and B, Neutral to alert attentive positions. C, Play-soliciting bow. D and E, Active and passive submissive greeting; note tail wag, shift in ear position, and shift in distribution of weight on fore and hind limbs. I, Passive submission with J, rolling over and presentation of inguinal-genital region. F and H, Gradual shift from aggressive display to ambivalent fear-defensive-aggressive posture. (Reprinted by permission of Coward, McCann and Geoghegan, Inc., and Blassingame, McCauley, and Wood from Understanding Your Dog by Michael W. Fox. Copyright © 1974 by Michael W. Fox.)

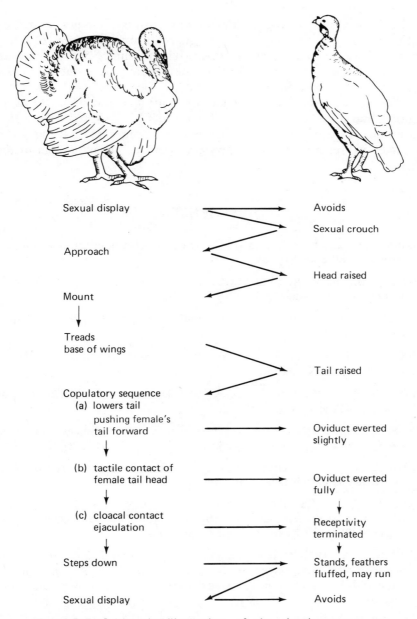

Male		Female
Sexual display	→	Avoids
		Sexual crouch
Approach	←	
		Head raised
Mount	←	
↓		
Treads base of wings	→	
		Tail raised
Copulatory sequence (a) lowers tail pushing female's tail forward	←	
	→	Oviduct everted slightly
(b) tactile contact of female tail head	→	Oviduct everted fully
		↓
(c) cloacal contact ejaculation	→	Receptivity terminated
↓		↓
Steps down	→	Stands, feathers fluffed, may run
Sexual display	← →	Avoids

Figure 5-5 Schematic illustration of the signal–response sequence of behaviors in turkeys required of the male (left) and female (right) for the successful completion of mating. (From Hale, Schleidt, and Schein, 1969.)

vocalizations, such as the boar's courting grunts. Females approaching estrus may be accompanied or "guarded" by the male for a day or two prior to their receptive stage. When females of the domesticated species are sexually receptive, they usually seek out a sexually active male if the pairing has not previously occurred.

The signal–response sequence of mating behavior patterns in turkeys has been studied by Schein and Hale (1965). Figure 5-5 shows the sequence of events leading to a completed mating. Each behavior is "released," in turn, by the previous behavioral act of the partner.

REFERENCES

DOMM, L. V., and D. E. DAVIS, 1948. Sexual behavior in intersexual domestic fowl. Physiol. Zool. 21:14-31.

FEIST, J. D., and D. R. McCULLOUGH, 1975. Reproduction in feral horses. J. Reprod. Fertil., Suppl. 23:13-18.

——, 1976. Behavior patterns and communication in feral horses. Z. Tierpsychol. 41:337-371.

FOX, M. W., 1974. Understanding Your Dog. Coward, McCann and Geoghegan, Inc., Publishers, New York.

GEIST, V., 1971. Mountain Sheep. A Study in Behavior and Evolution. University of Chicago Press, Chicago.

GROSSE, A. E., and J. V. CRAIG, 1960. Sexual maturity of males representing twelve strains of six breeds of chickens. Poult. Sci. 39:164-172.

GUHL, A. M., and L. L. ORTMAN, 1953. Visual patterns in the recognition of individuals among chickens. Condor 55:287-298.

HALE, E. B., W. M. SCHLEIDT, and M. W. SCHEIN, 1969. The behavior of turkeys. *In* E. S. E. Hafez (Ed.), The Behaviour of Domestic Animals, 2nd ed. The Williams & Wilkins Company, Baltimore.

HARPER, L. V., 1970. Ontogenetic and phylogenetic functions in the parent-offspring relationship in mammals. *In* D. S. Lehrman, R. A. Hinde, and E. Shaw (Eds.)., Advances in the Study of Behavior, Vol. 3. Academic Press, Inc., New York.

HEDIGER, H., 1950. Wild Animals in Captivity. Butterworths Scientific Publications, London.

RAMSEY, A. O., 1951. Familial recognition in domestic birds. Auk 68:1-16.

SCHEIN, M. W., and E. B. HALE, 1965. Stimuli eliciting sexual behavior. *In* F. A. Beach (Ed.), Sex and Behavior. John Wiley & Sons, Inc., New York.

SIGNORET, J. P., 1970. Sexual behaviour patterns in female domestic pigs (*Sus scrofa* L.) reared in isolation from males. Anim. Behav. 18:165–168.

TINBERGEN, N., 1965. Animal Behavior. Time-Life Books, New York.

6

Cyclic Activity and Systems of Behavior

Levels of arousal range from deep sleep to high levels of excitement and irritability. The skilled observer watching an animal during its sleep notes periodic changes in muscle tone, respiration rate, eye movements behind the closed lids, and facial and limb muscle movements. When electrical leads are placed on the skull above certain brain areas and attached to recording apparatus, characteristic wave patterns can be detected. Although it cannot be known directly whether animals dream, many who have observed dogs closely during sleep believe that they do. They often show muscular movements such as tail wagging and facial grimaces or vocalize by barking or whining while sleeping, as if experiencing a dream. During such episodes, their brain-wave pattern (electroencephalogram or EEG) resembles that of human subjects who are dreaming.

STATES OF ACTIVATION

Although level of arousal or sensitivity to stimulation exists along a continuum, there appear to be subdivisions that are associated with typical brain-wave patterns (e.g., Johnson, 1975). Figure 6-1 illustrates typical associations that have been found. Although the different stages of activation or arousal go by a variety of names, they can be characterized as follows.

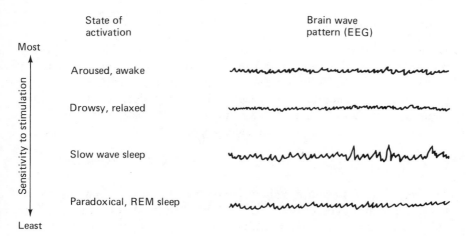

Figure 6-1 Sensitivity to stimulation varies from the condition of being fully aroused to that of deep sleep. Particular brain-wave patterns, as indicated by the electroencephalogram or EEG, are associated with the states of being awake, drowsy, "slow-wave sleep," and paradoxical or REM sleep. The deepest level of sleep is paradoxical in that the brain waves resemble those seen while individuals are fully awake and rapid eye movements (REM) occur, apparently during episodes of dreaming.

Rapid Eye Movement (REM) or Paradoxical Sleep

This stage has been demonstrated in many animals, including cats, rabbits, pigs, cattle, sheep, horses, dogs, pigeons, and chickens. The EEG pattern is rapid, of low amplitude, and closely resembles that seen in the animal while awake. It is paradoxical in that this is the deeper level of sleep (awakening the animal is more difficult) and yet the EEG corresponds to that of an animal already fully awake. Rapid eye movements, under the closed lids, occur during this kind of sleep and are an easily identified indicator of dreaming in human beings.

REM or paradoxical sleep is associated with reduced muscle tone, the postural muscles are lax, respiration and heart rates are lowest, and the animal is lying down. Nevertheless, it is during this phase that the facial and limb muscles cause occasional movements.

Slow-Wave Sleep

During this type of sleep, there is more postural muscle tone, bodily movements are lacking, and dreaming does not occur (at least in man). It seems to be present in all mammals and birds. Horses, and probably other ungulates, can be in slow-wave sleep while standing. Generally speaking, there are several periodic shifts between active and quiet sleep during a typical sleeping period.

Drowsy or Relaxed

The terms used here are self-explanatory. The animal may appear to be asleep with eyes closed, but the EEG pattern shows a very low amplitude, steady, and rapid pattern that is described as having "alpha waves." The animal is easily aroused into a fully active state, especially by sound.

Aroused or Awake

This stage covers a wide range of sensitivity to stimulation. It is common to include the drowsy state and extends to the condition of strong arousal or extreme irritability. Moderate activation is associated with turning the head or body toward stimuli, increases in muscle tone, heart rate, and respiration and is characterized by such behavior as moving about, feeding, and investigation. Strong arousal is associated with vigorous play, sexual, aggressive, and avoidance behavior.

SLEEPING CYCLES

Why do animals show regular, cyclic patterns of activity? Sleep tends to occur in every 24-hour period on a relatively regular basis and is clearly associated with daylight or dark periods. When daytime or nocturnal animals are mentioned, reference is to the period when they are normally active. Regular rhythms associated with 24-hour days are called *diurnal.*

Resting or restorative processes needed to bring the body back into a stable or homeostatic state after activity may help to explain the phenomenon of sleep, but this is clearly not the entire answer. An animal that has been exercised vigorously does not ordinarily restore its physiological state by sleeping immediately after, but continues in the waking state until that time of day when sleep is characteristic. Similarly, the amount of sleep and timing of sleep during 24-hour periods is relatively constant, even though level of activity may vary considerably from day to day.

Some sort of biological clock appears to be operating in maintaining the diurnal rhythm. It is suggested that sleep is adaptive behavior serving two primary functions. The first may be to reduce energy requirements during that part of the day when the animal is less suited to activity (because of specialization to a daytime or nocturnal life-style). The second may be to reduce exposure to predators, as sleeping animals attract less attention.

Ruckebusch (1972) conducted studies in diurnal cycles of activity of docile animals kept in barn stalls (horses and cows) or in metabolic cages (sheep and pigs). Brain waves were recorded directly from the cortex. Heart and respiration rates were measured and visual records were kept. Horses and sheep spent 90% and 80% of each 24 hours standing, whereas cattle and pigs stood only 40% and 20% of the time (figures are rounded to the nearest 5%). It has been noted that horses introduced into a strange environment commonly stand for several days. Ruckebusch also notes that animals tend to sleep less on pasture than when kept in familiar buildings. The amounts of time spent at different levels of arousal are given in Table 6-1 (rounded to nearest hour).

From the table it is evident that domesticated herbivores (cattle, sheep, and horses) spend most of their time in the waking state, although cattle are drowsy about 8 hours per day. Of approximately 4 hours spent sleeping daily, less than 1 hour is in the paradoxical or REM stage. In contrast, pigs, which are omnivores (eating both plant and animal foods), spent one-third of each day asleep under these conditions and had REM sleep (were they dreaming?) for about one-fourth of the sleeping period. For comparison, man (not shown

Table 6-1 *Hours at different levels of arousal during 24-hour intervals*

Animal	Awake			Sleeping		
	Drowsy	Moderate to strong arousal	Total	Slow-wave	REM	Total
Cattle	8	12	20	3	<1	4
Sheep	4	16	20	3	<1	4
Horses	2	18	20	3	<1	4
Swine	5	11	16	6	2	8

SOURCE: Data from Ruckebusch (1972).

in the table), also an omnivore, spends about the same number of hours daily in sleep and also has about 2 hours of REM sleep daily, as do swine.

SYSTEMS OF BEHAVIOR

Most goal-directed behaviors (such as feeding, drinking, sleeping, and sexual) consist of three distinct phases and occur in cyclic fashion. Because the phases were first studied with reference to food getting and feeding, two of the three were named accordingly. Thus, we speak of appetitive, consummatory, and refractory behavior. The appetitive phase may be simple or complex, but often includes searching and variable behavior and much that is learned. Consummatory acts tend to be relatively consistent or stereotyped, showing little variability from one individual to another, and may be largely instinctive. The refractory phase includes loss of interest and cessation of consummatory activity, even in the presence of continuing opportunity to respond.

Cycles may vary in length, and intervals between recurrence of the same phase may be irregular. For example, grazing animals may engage in several bouts of feeding activity during a 24-hour period, but the frequency and duration of grazing is likely to be influenced by daylight or darkness, temperature, nutritive value of forage, and the like. Carnivores, such as dogs and cats, when living in the feral state may feed at irregular intervals, with some feedings being separated by several days.

Some kinds of behavior are not necessarily cyclic, but occur as needed, and may not include all three phases. Protective reflexes, in particular, may lack appetitive and refractory phases. An animal will respond to a painful stimulus (such as a pinprick) by jerking away the injured part. No preparatory movements are required, the action is nearly instantaneous, and the response will recur again and again unless muscular fatigue interferes.

Functional Systems

Scott (1972:16-22), among others, has devised a classification scheme for the major kinds of behavior which has proved to be useful. As in any classification system, there are occasional difficulties in deciding to which category certain items should be assigned. Another problem is that complex behaviors, such as feeding or mating in competitive situations, may include more than one system.

Scott's categories are based on functions related to the animal's well-being.

Behaviorists often discuss patterns of behavior which are characteristic of particular species. There may be many patterns included within a system. For example, courting is a component of male sexual behavior, but the patterns of courting exhibited by turkey toms, boars, and bulls differ greatly.

Table 6-2 presents a list of the major systems of behavior, with examples to illustrate the three phases. Ingestive, eliminative, shelter seeking, and investigatory behavior may involve single animals; the remainder are social behaviors (except for self-grooming) in that two or more individuals are required for their expression. Each system will now be considered briefly, and all will receive more detailed study in later chapters.

Ingestive. Ingestive behavior includes both eating and drinking. Suckling activity of young mammals is included and usually occurs with little or no assistance other than the dam standing or lying quietly so that her offspring may reach the mammary glands. The appetitive behavior required for herbivorous, omnivorous, and carnivorous animals varies greatly. Grazing animals on pasture prepare for eating by simply lowering the head. Drinking requires travel to a source of water and the location of a watering site needs to be learned. Omnivorous and carnivorous animals tend to show much exploratory behavior in finding foods. Methods of obtaining food (especially by hunting prey) and food preferences are likely to be learned, to some extent, from one or both parents.

Eliminative. Elimination of urine and feces may be haphazard or controlled, depending on the life-style of the species. Those animals whose young are poorly developed at birth (altricial) tend to control their elimination and avoid contamination of the area where the young are kept, and they are able to learn with relative ease to avoid elimination in particular areas when trained by man.

Territorial and home range marking by dogs and cats involves urination on conspicuous objects within the area. The spilling of urine carrying certain odorous substances (pheromones) is typical behavior of adult mammals when sexually aroused and signals readiness to mate. These examples indicate some of the difficulties of classifying behavior as to system. Although all involve elimination, it is evident that avoidance of elimination in a den area may be related to care giving and predator avoidance for the young; territorial marking is related to agonistic and sexual activity, and the

Table 6-2 *Functional systems of behavior, with examples*

System	Example		Phases	
		Appetitive	Consummatory	Refractory
Ingestive	Horse grazing	Lowers head	Eats (long periods)	Rests and digests
	Cat with prey	Searching, stalking, attack, killing bite	Eats (short periods)	Rests, may guard or hide remnants or carry to young
Eliminative	Cow	Elevates tail, "hunches" back	Eliminates	Ignores waste
	Cat	Sniffs, selects site, elevates tail, etc.	Eliminates	Scratches material to cover waste
Shelter seeking				
Resting	Dog	Selects site, circles, lies down	Rests or sleeps	Stretches, becomes active
Predator avoidance	Chicken	Orients to predator, gives warning call (especially cocks with flock or hen with chicks)	Immobility or active flight to refuge	Alertness decreases gradually as predator moves away, normal activity resumes
Investigatory	Horse and novel object	Approaches cautiously; may run off briefly, then return	Looks, listens, smells, may touch and taste	Moves away, loses interest
Allelomimetic	Wolves hunting	Howl, come together, social greetings are exchanged	Cooperative hunting and killing of large prey	Feeding, gradual loss of contact between group members
	Cattle remaining in herd	Become agitated if separated (especially if predators are present)	Rejoin group	Recover from anxiety, resume normal activities
Agonistic	Strange dogs	Approach alertly, attempt identification, especially by odor	Aggressive: threatens, bites	Dominant "stands over," loses interest
			Submissive: lowering of head, tail between legs, rolling over	

Table 6-2 *(continued)*

		Phases		
System	Example	Appetitive	Consummatory	Refractory
Sexual	Bull mating with cows	Moves among cows, observes postures, smells, nudges cows, mounts cow, inserts penis	Ejaculates	Loses interest temporarily
Care-giving (epimeletic) Self-care	Cat grooming		Licks fur and paws, rubs against objects	
Care of young	Mare with newborn foal	Turns to foal, bends neck	Licks, cleans off fetal membranes, stimulates foal to activity, stands for nursing	Walks away
Care-seeking (et-epimeletic)	Chick separated from dam or group	Becomes aroused, looks, listens	Gives "distress" call	Becomes quiet after "rescued" by mother hen, may give "contentment" chirp

release of pheromones in the urine can lead to or be a part of sexual behavior.

Shelter Seeking. Locating and making use of safe or comfortable shelter and avoiding dangerous or unhealthy situations are included under shelter seeking. When a dog walks to a suitable location, sniffs about, circles, and lies down, he is showing appetitive behavior under this system. Predator avoidance is also included, and higher animals may show altruistic behavior (they may reduce their own safety by warning or protecting others) when a predator's presence is detected. Thus, McBride et al. (1969) noted how feral cocks remain alert while their hens and chicks feed. When a cat was observed approaching stealthily, the cock would give a specific call associated with the presence of a ground predator, and all members of the group would then remain alert until it moved away. Animals in a group are generally less susceptible to predator attack than is a solitary animal. Other animals' bodies in many situations provide some degree of shelter or increase comfort (Figure 6-2).

Figure 6-2 Baby pigs benefit from shelter and comfort obtained by maintaining close bodily contact in a chilly environment. This behavior may be classified as shelter-seeking. (Photograph by the author.)

Investigatory. Investigatory behavior may be relatively subtle. Many grazing animals will frequently raise their heads to look about, listen, sniff the air briefly, and then return to feeding. Vision, hearing, and olfaction (smelling) are all useful for maintaining an awareness of the ongoing situation and are particularly evident among animals that were likely to be preyed upon prior to domestication. Other senses come into play with closer investigation.

When a horse becomes aware of a strange object, it is likely to examine it at a distance by the senses already mentioned. If the object is not frightening, the animal may then approach cautiously, a few steps at a time; it may run away briefly and then return. Eventually, the object may be touched and perhaps licked or tasted as well. Satisfied by its investigation, the horse will eventually lose interest and move away.

Allelomimetic. Allelomimetic behavior includes a wide variety of activities in which animals do the same thing or cooperate in some fashion. Mutual stimulation is a common feature of such behavior.

Different functions may be served by allelomimetic behavior. Seeking shelter by joining and staying with a group involves allelomimetic behavior. The interrelationship of allelomimetic and shelter-seeking behavior is evident from Francis Galton's lively description in 1871 of cattle exposed to lions in South Africa (quoted by Wilson, 1975:38): "Yet although the ox had so little affection for, or individual interest in, his fellows, he cannot endure even a momentary severance from his herd. If he be separated from it by strategem or force, he exhibits every sign of mental agony; he strives with all his might to get back again and when he succeeds, he plunges into its middle, to bathe his whole body with the comfort of close companionship."

Social facilitation is a form of allelomimetic behavior in which the beginning of an activity or increased activity by one individual leads to similar behavior of others. It is often associated with feeding, and as indicated later, may be an important means of increasing food intake.

Agonistic. Agonistic behavior embraces both aggressive and submissive acts, including escape of a subordinate animal from a dominant one. Although this kind of behavior may be undesirable if carried too far, it serves in animal societies to organize groups and to maintain social hierarchies. Organized groups are better adapted, in general, than are disorganized ones.

Sexual. Males and females have very different roles in sexual behavior among polygynous species. The male shows a great deal of courting and seeking out of receptive females. If not located in early estrus by a male, the female may, when in full heat, search for an active breeding male. A complex ritual or sequence of events then occurs in which there is a chain of cause-and-effect relationships, which culminate in insemination and a subsequent refractory period.

Care-giving (Epimeletic). Generally, care-giving is regarded as maternal behavior, but self-care is also included (as when a cat cleans itself or a hen preens her feathers). Maternal care is of major importance in those animals where the young are born in the open and are susceptible to adverse weather conditions or predation. Parental care is also essential for species that are poorly developed at birth (e.g., cats and dogs). A great deal must be learned about care giving when, for management reasons, the young are removed from the mother at ages before weaning or separation would occur naturally.

Care-seeking (Et-epimeletic). In a sense, care-seeking and shelter-seeking behavior are related systems when the presence of other animals of the social group provide protection or increase well-being. The care-seeking category will be limited here to interactions with other animals. Communication of need for care often requires vocalization or tactile stimulation (as when a pig or calf nuzzles its mother's udder).

Other Systems. Although other functional systems have been proposed, such as leadership and territoriality, it appears that they have considerable overlap with those already listed. Leadership may be considered under allelomimetic behavior, and territoriality, which involves defense of an area, includes aggressive behavior. Most behaviorists are uncertain of the function of play. It may be part of investigatory behavior, in which the animal learns what it is capable of doing and develops synchrony of movements necessary for smooth functioning of behavioral patterns when those are needed later in life.

Displacement activities are ordinary but inappropriate or out-of-context behaviors. They are most likely to be seen when strongly conflicting drives are present or when an animal is frustrated or stressed. An example, often noted by those observing initial encounters between strange chickens, is that if agonistic behavior continues for some time it may be abruptly interrupted by "tidbitting." Following a vigorous bout of fighting, one or both individuals may peck or scratch at the bare floor as if grain or other food were present. The fight may then cease entirely or may begin again. Other examples include hens kept in crowded cages, which spend apparently excessive amounts of time preening their feathers, and sheep isolated from their flockmates in barren enclosures, which appear to be grazing although no forage is present.

REFERENCES

JOHNSON, J. I. Jr., 1975. States of activation: sleep, arousal and exploration. *In* E. S. E. Hafez (Ed.), The Behaviour of Domestic Animals, 3rd ed. The Williams & Wilkins Company, Baltimore.

McBRIDE, G., I. P. PARER, and F. FOENANDER, 1969. The social organization and behaviour of the feral domestic fowl. Anim. Behav. Monogr. 2:127–181.

RUCKEBUSCH, Y., 1972. The relevance of drowsiness in the circadian cycle of farm animals. Anim. Behav. 20:637–643.

SCOTT, J. P., 1972. Animal Behavior, 2nd ed. University of Chicago Press, Chicago.

WILSON, E. O., 1975. Sociobiology. Belknap Press of Harvard University Press, Cambridge, Mass.

7

Genetics of Behavior

GENETIC VARIATION

When any species that has been under man's control for a few thousand years is considered, it is evident that distinctive types (and breeds within types) have been developed to fit those environmental niches under man's domain. Occasionally, nature may deny the use of a species in a particular place and use her own choices instead. (Consider our inability to substitute cattle for other ungulates inhabiting large areas of Africa where the tsetse fly and sleeping sickness occur.) Nevertheless, man's success in sorting out breeds within species that possess those shapes, functions, and behavioral traits desired for particular purposes is impressive. Much of that success was achieved by selection, with only limited knowledge of the laws of heredity. In other cases, entire subpopulations were substituted for others after they had become different because of relative fixation of desired traits in small, genetically isolated populations.

Within breeds there may be varieties or strains that differ substantially, in behavior, and within those, family differences are recognized. Full brothers and sisters also differ when traits are measured carefully. In fact, even identical twins and members of highly inbred lines, which have the same genetic constitution, may show differences, especially if they have been deliberately exposed to different social environments. This last observation reminds us that behavioral (and other) characteristics are determined not only by heredity, but also by the internal and external environment to which they are exposed, as indicated in Chapter 1.

SEXUAL EFFECTS

Some of the most profound differences in behavior are associated with sex and with changes resulting from loss of sex by castration. Those will be dealt with primarily in later chapters. Why are there two sexes in the first place? Sex obviously has to do with reproduction, and yet many species of the plant kingdom typically reproduce or may be artificially propagated by asexual processes. With asexual reproduction, large numbers of individuals can be produced that are genetically identical. This situation is favorable if the environment is relatively stable and the species is adapted to it. If the environment changes, the organism may no longer be well adapted. Even so, an organism with great reproductive power (such as bacteria) may persist if occasional genetic changes (mutations) occur. Mutant forms take over and cope with the altered environment.

Sexual reproduction is characteristic of life forms that are generally capable of doing well in a wide variety of habitats or in environments that fluctuate considerably. Nevertheless, a price is paid for sexual reproduction (Daly and Wilson, 1978:38–47); a major part of that price is the production of some individuals that are inferior. The primary biological function of sex and of sexual reproduction is to allow fresh samples to be drawn from the population's gene pool each generation, thereby maintaining genetic variability.

SEX DETERMINATION

Domesticated animals are genetic diploids. That is, each individual has within the nucleus of each cell two sets of paired chromosomes, one member of each pair being received from the mother and the other from the father. Paired chromosomes that carry genes are called *autosomes*. The genes or hereditary factors determine the kind of animal that develops after fertilization and influence details of its phenotype, together with the environment. All chromosomes are autosomes except for one odd pair, the *sex chromosomes*.

The sex chromosomes are of special interest because the pair is alike in about half of all individuals but unlike in the other half. When they look alike microscopically, both carry many pairs of genes, one gene of each pair being received from each parent. When unlike, one member carries genes but the other member does not, or it may carry only a few and those are not like the genes of the fully endowed member. Among mammals, the gene-carrying sex chromosome is called the X and the essentially blank (and smaller) sex chromosome is the Y. Genes carried on the X are called sex-linked, although it would seem more logical to speak of them as X-linked. Perhaps the most astounding thing about this pair of chromosomes is that somehow it controls sex; individuals carrying two X's (XX) are female and those with one X and one Y (XY) are male. Birds differ in sex determination from mammals because in birds it is the male that has the like pair of sex chromosomes (usually designated ZZ) and the female that has the unlike pair (ZW).

The chromosomal content of the reproductive cells (sperm and ova) and of the fertilized ova from which individuals develop are shown diagrammatically in Figure 7-1 for both mammals and birds. In mammals, the dam always contributes an X chromosome to her

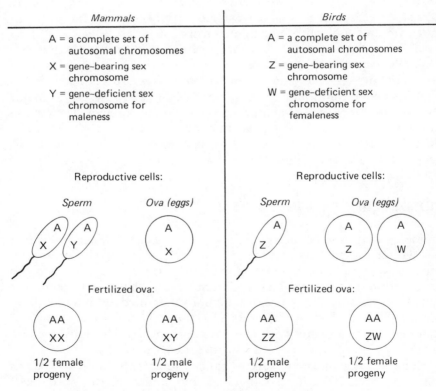

Figure 7-1 Diagrammatic illustration of the genetic content of male and female somatic and reproductive cells (gametes) and of sexual determination in mammals and birds.

progeny. Sex is therefore determined by whether the particular sperm that fertilizes an egg is X- or Y-bearing. The figure shows why essentially half of all progeny produced in a population should be of each sex, although the exact ratio may deviate slightly from that (it is about 105 males:100 females at birth in man). It is also apparent that all X-linked genes of male progeny are received from the mother in mammals, and all Z-linked genes of daughters are contributed by the sire in birds.

SEX-LINKED EFFECTS

At times it becomes evident that specific sex-linked genes or perhaps several genes on the X or Z chromosome influence a behavioral trait. Several decades ago, breed-crossing studies in chickens indicated that the breed of the sire had a greater effect on the incidence of broodiness in hens than did the breed of the dam.

White Leghorn chickens have been selected to be nonbroody, but some other breeds continue to have a high incidence of broody hens. If White Leghorns (WL) are crossed with a strain of another breed having broodiness (e.g., Cornish), the following results would be expected:

Cross of parental stocks	Phenotype of F_1 hens
WL ♂ X Cornish ♀	Nonbroody
Cornish ♂ X WL ♀	Broody

The daughters show the behavior characteristic of hens of the breed of their sires. As Figure 7–1 illustrates, the sex chromosome of avian males are alike and designated ZZ, but females have the unlike pair ZW. If it is assumed that the Z chromosome has most of the genes influencing broodiness, the results can be explained. Let the White Leghorn Z chromosomes be labeled Z^-, with the $^-$ superscript indicating nonbroody alleles and the Cornish Z as Z^+, indicating the presence of alleles for broodiness. The example is then reconstructed as follows:

Genotypes of parental stocks		Genotypes of F_1 hens
WL ♂ X Cornish ♀		
Z^-Z^- Z^+W		Z^-W (nonbroody)
Cornish ♂ X WL ♀		
Z^+Z^+ Z^-W		Z^+W (broody)

It is clearly evident that broody behavior of hens is influenced by one or more Z-linked or sex-linked genes. Why do males of strains in which hens show broodiness not also show this kind of behavior? The answer is found in the internal environment, which differs between male and female. If cocks of a strain with broody hens are injected with high levels of prolactin, they also show broody behavior. It may then be stated that broodiness is influenced by sex-linked genes, and, in addition, is a sex-influenced trait normally observed in females only.

MATERNAL EFFECTS

When the progeny of reciprocal crosses between parental stocks that differ in a behavioral trait are compared, there is a need to consider the possibility that differences between the F_1 progeny of the two crosses may be caused by differences in the maternal environment either before or after birth. Some experimental results obtained by Scott and Fuller (1965) involving reciprocal crosses of the Cocker Spaniel and Basenji breeds of dogs are informative here. Cockers are relatively nonaggressive and confident with people, as contrasted with Basenjis, which are highly aggressive and wary. After the reciprocal crosses were made, F_1 progenies were compared for behavioral traits. For most behaviors there was no evidence of a difference between the crosses. However, for attraction behavior, scored as the number of times 13- to 15-week-old pups approached or followed the experimenter, the F_1's behaved in the same manner as the breed of their mother. Similarly, F_2 pups, obtained by mating together F_1's of the same parental cross, showed the same behavior:

| | Attraction behavior score | |
Parental cross	F_1	F_2
Basenji ♂ X Cocker ♀	Similar to Cockers	Similar to Cockers
Cocker ♂ X Basenji ♀	Similar to Basenjis	Similar to Basenjis

It appears that approaching and following human beings must have been learned from the mother and persisted from weaning at about 8 weeks of age until the pups were tested at 13 to 15 weeks. Continuation of the behavior into F_2 pups indicates that this behavior persisted as a social tradition by way of maternal influence during the preweaning period.

From the examples it may be seen that sex-linked effects are likely when reciprocally crossed breeds produce F_1 progenies that differ in one sex only; the other sex (receiving gene-bearing sex chromosomes from both parental stocks) should show no differences because of sex linkage. On the other hand, when both sexes of the same F_1 cross resemble the female parent stock, maternal influence is likely.

SIMPLY INHERITED TRAITS

When different forms of the same gene exist, the alternative forms are called *alleles*. When snorter dwarfs of cattle were considered in Chapter 3, it was noted that the "normal" allele (Dw) was dominant, and that individuals carrying the recessive allele for dwarfism (dw) in heterozygous condition (Dwdw) were not dwarfs. Although Dwdw individuals are rather blocky or compact in body type, they are within the normal range of variation. The abnormal allele dw presumably arose because of a rare and random change (mutation) in the biochemical structure of the gene.

Genes, for which there are two or more allelic forms, such as those for normal and dwarf body size, having obvious effects are called *major* genes. Characteristics influenced by major genes are simply inherited and are termed *qualitative traits,* as individuals can easily be sorted into alternative groups on the basis of their phenotype. In the case of cattle, snorter dwarfs and nondwarf individuals can easily be classified by simply looking at the animals.

Dwarfism in Sheep: An Example

The allele for achondroplastic dwarfism (shortened long bones) is recessive in cattle and sheep. Sheep with extremely short legs, but otherwise appearing relatively normal in their other body parts, "breed true" when mated together. At one time they were favored by some farmers, who kept them on relatively lush pastures, where they were not required to walk long distances for grazing or water. Low fences were adequate for keeping such sheep confined and the Ancon breed was formed.

If the designation of "an" is used for the allele causing this kind of dwarfism in sheep, the dwarf genotype may be designated anan, indicating that such animals are homozygous recessive. Those alleles occur on autosomal chromosomes so that both male and female dwarfs have the same genotype. Because an is recessive to the normal allele An, heterozygous individuals (Anan) appear normal as the recessive gene is hidden; the heterozygous individual in such a case is often called a *carrier*. Individuals having a pair of normal alleles (AnAn) at this chromosomal location or locus would be called *homozygous dominant* or *normal*. When sperm or eggs are formed, one member of each pair of autosomal genes at the same locus goes into each gamete. Heterozygous individuals would, therefore, form equal numbers of reproductive cells having either the An or an allele. Homozygous individuals would have all reproductive cells alike for genes at this locus; AnAn parents would produce gametes alike in that all carried an An allele, and anan parents would have gametes carrying an.

With this background, the outcome of various kinds of mating can be predicted. Because the an mutation is rare, the probability is very high that most individuals outside the Ancon breed are homozygous normal (AnAn). Therefore, if an achondroplastic dwarf sheep (anan) is found and mated with an unrelated animal (AnAn), the phenotypic and genotypic results will be:

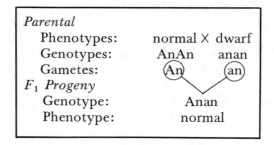

What would the result be if such F_1 progeny were mated? This would be the typical approach of a breeder who has what appears to be a "new" genetic mutation and wants to use it to develop a true-breeding strain. The situation would then be:

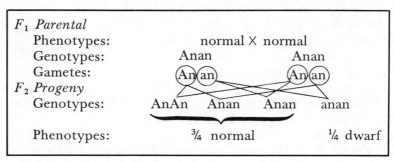

The 3 : 1 ratio of normal to abnormal in the F_2 progeny is the typical ratio expected. Because chance is involved as regards which gametes actually come together at fertilization, the ratio may deviate from the expected when only a few progeny are produced. Nevertheless, if a sizable number of offspring are born from such matings between F_1 parents, the typical ratio should be approached closely.

As soon as two homozygous recessive dwarf sheep are produced (of opposite sex!), a breeder can proceed to produce more dwarfs, without exception, by mating dwarfs together.

General

Mutations causing extreme phenotypes are not necessarily deleterious. Nevertheless, it has often been observed that any extreme phenotype produced by a mutation (either dominant or recessive) at a single locus is likely to have multiple effects, and there are frequently unfavorable physiological changes which reduce the fitness of individuals exhibiting the extreme phenotype.

Although mutations having large effects are favored for illustrating genetic principles (as above), it is important to realize that relatively few behavioral traits are simply inherited (involving only one or a few gene pairs). More commonly, genes at several or many loci (polygenes) act together in influencing a trait. Nevertheless, the mechanics of inheritance are basically the same, whether or not major genes or polygenes are involved.

INBREEDING EFFECTS

Intensive Inbreeding

When related individuals are mated, their progeny are inbred; they have fewer heterozygous gene pairs than if the parents had not been related. If the exact genotype of every individual in a population were known for the thousands of pairs of genes involved, and an inbreeding study were then carried out, the average loss of heterozygosity (or increase in homozygosity) could be followed: As an example, consider that the average animal in a particular population has 1000 pairs of genes in which alleles differ. The most rapid loss of genetic heterozygosity (conversion of heterozygous to homozygous gene pairs) in animal populations occurs when full brothers and sisters or parent and offspring are mated to produce young. With continuing brother–sister matings in the hypothetical population,

the number of heterozygous loci would be expected to decrease in
the following manner:

Number of generations of brother–sister matings	Heterozygosity lost (inbreeding coefficient)	Number of heterozygous loci remaining
0 (original population)	0	1000
1	0.250	750
2	0.375	625
3	0.500	500
4	0.594	406
5	0.672	328
.	.	.
.	.	.
.	.	.
10	0.886	114

Besides the general loss of heterozygosity, other things would be
happening as inbreeding progressed. The population would become
subdivided into many small subpopulations, as each brother and
sister pair would be the basis of a separate "line." Matings would be
restricted so that only the closest relatives would be allowed to
mate (matings between lines would be prohibited). The different
lines would "drift" apart from each other genetically as different
combinations of alleles became homozygous at the various loci in
each line, but individuals within lines would become more alike.

Because alleles having less favorable physiological effects are
usually recessive, a loss of vigor would result as inbreeding progressed;
physiologically unfavorable effects would be expressed as more and
more genetic loci with recessive alleles became homozygous. For
some gene pairs, it is known that the heterozygous condition produces a more desirable phenotype than either homozygous combination; for example, Aa could be superior phenotypically to both
AA and aa (called *overdominance*). Thus, inbreeding would lower
the phenotypic value of the population in the case of gene pairs with
overdominant effects.

Briefly stated, close inbreeding in a population would have these
effects:

1. The population would be divided into subpopulations or lines.

2. The lines would become quite different from each other.

3. Uniformity would increase within the lines.

4. There would be a general loss of vigor.

Among domesticated species of animals, inbreeding in such drastic fashion (by mating brothers and sisters over several successive generations) has been limited to a few small, experimental populations. An early study was reported by Hodgson (1935), who began with a sample of the Poland China breed of swine. After eight generations of matings between full brother and sister pairs, only three of eight original lines had survived, and only two of those (M and N) were present in sufficient numbers to compare. The expected loss of vigor had occurred with a vengeance! Along with delayed maturity, reduced disease resistance, and smaller litter size, he found that the sex drive of the boars was greatly reduced. Hodgson was impressed by the behavioral differences between lines, particularly by differences in maternal care and temperament. Both the M and N lines had relatively high mortality of baby pigs, but for very different reasons. Line M sows and pigs were of mild temperament; the pigs were described as "friendly" and the sows were quiet. They would allow a child to play with their piglets. Baby pig loss in line M was often caused by sows lying on their pigs; they were fairly indifferent to the squeals of the pigs that were laid on. In contrast, line N animals were "wild." They were savage in their defense of baby pigs and often trampled their own young in attempting to defend them from trivial disturbances.

Highly inbred lines of mice, rats, guinea pigs, and chickens have also been produced. Although behavioral traits have not usually been the focus of those studies, it is a common observation that each line has its own peculiar set of behavioral characteristics, even when those lines were derived from the same base population.

Moderate Inbreeding With Selection

Because of the drastic loss of vigor commonly encountered with intensive inbreeding, most livestock breeders avoid mating relatives. Nevertheless, the increased uniformity obtained and the occasional fixation of desirable combinations of genes within lines suggest that there may be some value in inbreeding if reproductive capacity is not of major importance. When skillful selection is combined with moderate inbreeding (as in mating cousins, uncles and nieces, etc.), anatomical and behavioral traits may respond favorably. Thus, breeders of "show" animals often engage in the practice of combining selection with mild inbreeding.

Desirable results have been achieved using this approach in the breeding of guide dogs for the blind (Pfaffenberger et al., 1976:

168-175). Although experiencing some loss in reproductive fitness and survival of baby pups, three moderately inbred lines of German Shepherds were produced that had greater uniformity in appearance and trainability together with improved temperament. A typical animal produced in this way is shown in Figure 7-2.

HYBRIDIZATION AND HETEROSIS

Plant breeders make extensive use of hybrid vigor or heterosis when they cross inbred lines, as in the production of hybrid corn. Poultry breeders follow somewhat similar procedures in producing incross, straincross, and crossbred chickens, and large animal breeders commonly cross breeds of beef cattle, swine, and sheep. The benefits are most apparent in terms of fitness components. Growth rate usually increases, sexual maturity occurs earlier, there is greater productive capacity, and infant mortality is generally reduced among crossed progeny.

How would hybridization affect behavioral traits? When those traits influenced by polygenes are considered, two general kinds of response occur among first-cross (F_1) progeny:

1. Heterosis generally occurs for fitness and its components when the stocks crossed are members of the same species.

2. Crossed progeny are usually near the average of the parental

Figure 7-2 A typical German Shepherd bitch resulting from selection combined with mild inbreeding by Guide Dogs for the Blind, Inc. Among favorable behavioral changes resulting from this program are reduced walking speed more suitable for guide dogs than generally occurring in the breed, improved temperament, and greater uniformity in response to training. (From Pfaffenberger et al., 1976.)

stocks for nonfitness traits; thus, the length of long bones is so affected.

Guide-dog behavior might be improved by hybridization between carefully selected breeds or by crossing of lines within a breed. Pfaffenberger et al. (1976:212–215) recommended that experimental crosses be made between German Shepherds, Labradors, and Golden Retrievers for this purpose, as those three breeds have all proved useful as guide dogs for the blind. Reproductive performance and general vigor would surely be improved, and there is evidence from other breed crosses that hybrids may be exceptionally gifted with respect to trainability.

First-cross hybrids should be relatively uniform in all traits. Phenotypic uniformity is produced whenever individuals have the same genotype, even if many heterozygous gene pairs are present. That relatively heterozygous individuals can be phenotypically uniform is shown by the remarkable likeness of identical twins in man and in cattle and by hybrid corn. In fact, there is evidence that hybrids, produced by crossing inbred lines, often show more uniformity than do individuals within the parental lines. Heterozygous individuals appear to be better "buffered" than inbreds, so that minor environmental stresses that might produce adverse effects in inbreds have lesser or no effects in the hybrids. Crosses between breeds which themselves are relatively uniform should also produce vigorous and uniform hybrid progeny. Breeds within a species are equivalent (in comparative terms) to moderately inbred lines within a breed.

BEHAVIORAL TRAITS INFLUENCED BY POLYGENES

Animals usually differ from each other in relative terms rather than by being easily classified into categories, such as dwarf or normal. As an example, a great deal of variability is evident when differences are considered in the ease of training individual dogs or different breeds of dogs for simple acts of obedience, herding, guarding, retrieving, or hunting. Some individuals and breeds are better adapted to certain kinds of training or activity because of physical abilities (Greyhounds catch fast-running prey more effectively than do Poodles) or temperament (German Shepherds make better guard dogs than Collies).

Comparisons of relative performance or behavioral abilities are most easily made if such characteristics can be measured numerically.

In most cases individuals within populations vary in such a way that a bell-shaped frequency distribution can be plotted. There are relatively few extreme individuals at either end of the distribution and most tend to be somewhere near the average. Different populations, such as breeds within a species, tend to have similar distributions and those frequently overlap considerably, although average values may differ, as shown in Figure 7-3. Thus, although Greyhounds generally run faster than Poodles do, it may reasonably be expected that fast-running Poodles would outdistance some of the slower Greyhounds. In the figure, population A could represent the Poodle population and B the Greyhounds; the faster individuals within the Poodle population would be represented under the right-hand extreme of their distribution curve and the slower Greyhounds under the left-hand portion of their curve.

How can some understanding be gained of the relative importance of heredity and environment in shaping characteristics that vary so much and are best measured by numbers? The traditional Mendelian method of studying genetic influences is to cross relatively uniform, "true-breeding" populations that differ considerably and then examine their cross progeny (the F_1) and the progeny produced by mating F_1's together. This method has proven to be useful in studying traits, such as dwarf and normal, in which individuals are easily classified. However, that technique is not appropriate for studying populations that show considerable variability.

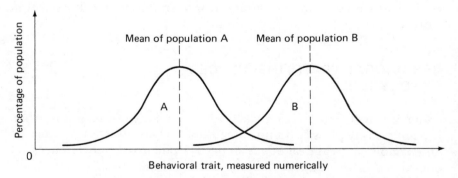

Figure 7-3 Hypothetical frequency distributions of two genetic populations (A and B) which have average values that differ, but with overlapping ranges. Most individuals within each population are relatively near the mean (represented by the vertical lines at the center of each population's distribution), while fewer and fewer individuals are found as deviations increase from the means of the respective populations.

If the average value of a characteristic of a breeding population can be changed by selection, there is clear evidence of genetic influence. Inbreeding depression and hybrid vigor also indicate genetic influences, but of a different kind. The kind of genetic variability (additive variance) will be considered here which causes a change when selection pressure is applied.

Effects of Selection: An Example

Can selection exerted by man change a behavioral trait within a few generations? The two studies to be considered provide valuable information and represent the best information obtainable under the circumstances. Nevertheless, the apparent responses to selection may have been overestimated, as the results can be interpreted in more than one way because of possible confounding of genetic and environmental effects.

The Royal Guide Dogs for the Blind Association of Australia appears to have produced remarkable improvement in a selected strain when compared with unselected stock over 4- and 5-year periods (Goddard and Beilharz, 1974). The Labrador breed was used primarily and pups were obtained from two sources, by donations from the public or from their own selected stock. Pups obtained from the public are obviously not obtained from parents selected for ability to lead the blind and would, therefore, appear to be suitable for comparisons with pups of the selected stock in order to estimate the success of the breeding program. Of pups donated by the public, only one-third eventually became guide dogs, but nearly two-thirds of selected stock pups reached the same goal. Less dramatic but similar results are suggested from comparisons of donated and selected-strain dogs of the German Shepherd breed in California (Pfaffenberger et al., 1976).

Should we accept the Australian and California results as indicating a large genetic effect on the ability of pups to qualify as guide dogs? Certainly, animals of the selected strains showed greater success. But donated pups were reared in different environments before being given to the training institutions. Close scrutiny of both programs reveals that the majority of those animals failing did so because of fearfulness in critical situations. Could it be that early experience might modify this behavioral characteristic? Maternal effects, prior to weaning, were shown to influence social attraction scores of crossbred pups when tested at a later age. Similarly, early experience of pups appears to modify the probability of fearful behavior being shown at later ages (Pfaffenberger and Scott, 1959).

Thus the kind of difficulty that can interfere with a clear understanding of the relative importance of genetic and environmental
influences is encountered. Carefully designed experiments can
reveal more about the relative effectiveness of selection in altering
behavioral traits.

Heritability and Effectiveness of Selection

The comparisons described above suggested, but did not prove, that
selection can be used successfully for increasing the ability of dogs
to lead the blind. However, those studies did not indicate, in a
quantitative way, the actual effectiveness of selection (realized
heritability). Heritability estimates can be obtained before selection
begins and are useful in indicating the probable rate of improvement
in proposed selection programs. When records of a number of families
are available, and those families have been kept under the same
general environmental conditions, relatively reliable estimates of
heritability can be made. Falconer (1960) describes such methods
of estimating heritability.

Heritability indicates the relative importance of polygenes as
compared to all sources of variability affecting a particular trait
within a generally specified environment. Heritability, in the "narrow
sense," indicates how effective selection should be. A specific example
should make this clear. Van Vleck (1977) presents a summary of
heritability estimates for some traits of horses. Those for running
speed range from 0.25 to 0.50. Conservatively, the heritability of
running speed in Quarter Horses may then be assumed to be about
0.25, or 25%. Thus, if horses are selected on the basis of their own
speed in running a quarter mile, progress should be made at the rate
of 25% of what is selected for in each generation. This estimate
indicates that, *on the average*, 25% of the superiority (or inferiority) of a particular animal is associated with genetic superiority
(or inferiority) and the other 75% is associated with environmental
effects over which the breeder has no control and cannot adjust for
by correcting for such variables as age, sex, and weight of the rider.
To be more specific, assume that the average running speed of a
group of prospective breeders is 40 miles per hour (mph). If parents
are selected that run at 44 mph, there is a selection differential of
4 mph. The expected genetic progress for one generation would
then be equal to heritability × selection differential, or, in this
example, 25% × 4 mph, which is 1 mph. Thus, the average running
speed of the next generation should be 41 mph as compared to the
40 mph of the previous generation.

Relatively few heritability estimates are available for behavioral traits; Table 7-1 presents some of the estimates. Most of the traits studied should respond to selection pressure. Appetite of broiler-type chickens, measured in terms of feed consumption, was estimated in 1958 as being highly heritable, and experience indicates clearly that such chickens have increased greatly in their feed consumption rates since then, presumably in response to selection for rapid gains in body weight, which requires a high level of feed intake.

In selecting for traits of relatively low heritability, records of closely related animals can increase the effectiveness of selection. This is likely to be useful if those relatives are kept in the same environment and tested at the same time. Thus, either full or half-brothers or -sisters are most likely to have comparable and reliable records.

Ancestors' records may be misleading in deciding which of several animals to keep as breeders if only a few ancestors' records are available and if those animals were kept and tested under different conditions. On the other hand, progeny testing may be a valuable tool (as in sire testing), because many records may be available and there is more likelihood that they may have been collected under specified conditions.

Although response to selection may be small in a single generation (when heritability is small or moderate), the genetic response is

Table 7-1 *Heritability estimates for some behavioral traits in domesticated animals*

Species	Behavioral trait	Heritability (%)	Source[a]
Cattle	Temperament (ease of handling during milking)	47–53	D
	Social dominance value	0–29	D
Swine	Avoidance learning (at 3 weeks of age)	50	W
Horse	Running speed	25–50	V
	Walking, trotting speed	40	V
	Points for movement	40	V
	Points for temperament	25	V
	Pulling power	25	V
Chicken	Feed consumption (broilers), 4–8 weeks	86–96	T
	Aggressiveness, social dominance	16–57	S
	Mating frequency of males	18, 31	S
	Learning factors	9–28	S

[a]Coded as follows: D, Dickson et al. (1970); S, Siegel (1975, review of literature); T, Thomas et al. (1958); V, Van Vleck (1977, review of literature); W, Willham et al. (1963).

cumulative and large phenotypic differences may be produced after several generations of selection.

Comparisons across generations, years, farms, and the like are often unsatisfactory for determining whether changes due to selection have actually occurred. Uncontrollable variables, such as weather, nutritional changes, disease exposure, social experiences, and difference in caretakers, will be confounded with the effects of genetic change. Scientific studies usually include unselected "control" populations being kept in the same environment. Successive comparisons of selected and unselected animals from such populations allow relatively unbiased estimates of selection effects.

Selection Studies

Although behavioral traits are often of major importance to large-animal breeders, controlled selection studies relating to behavioral traits have not been carried out with them (Siegel, 1975). However, selection studies involving learning ability in the rat and mating ability and agonistic behavior in chickens have been conducted. They will be reviewed briefly to indicate what is known about selection responses in those species and to give clues as to possible responses to be expected in other animals.

Learning Ability. Learning ability is of special interest to breeders and trainers of companion and working animals. Five separate studies, using laboratory strains of rats, concentrated on producing "bright" and "dull" strains by testing and bidirectional selection on performance in various types of mazes (summarized by Fuller and Thompson, 1960: 207-214). In four of the five studies, clear-cut differences were produced. Realized heritabilities were not calculated, but such large differences were present that bright and dull populations were essentially free of overlap after four to seven generations of selection. It is important to note that the changes in learning ability resulting from selection were rather specific and did not necessarily carry over to learning how to negotiate very different kinds of mazes. Thus, selection was effective in changing the ability to solve particular kinds of problems, but general intelligence was not necessarily altered.

Studies of learning ability in dogs have centered on comparisons of breeds in laboratory settings (Scott and Fuller, 1965). Although the role of selection in producing the differences observed is obscure, breeds differ in the ease with which they learn or can be trained to show different behaviors. For example, Basenjis and Wire-haired

Terriers are more readily trained to perform tasks or solve problems in ways requiring independent action, but Cocker Spaniels excel in obedience training.

Mating Frequency of Males. High levels of sex drive or libido and the related behavior of mating are important in all domesticated species. Ability to mate successfully with a large number of mates is of particular importance in the male. When artificial insemination is to be used, willingness and ability of the male to ejaculate following artificial stimulation is significant.

Siegel (1965) conducted a long-term selection study of mating ability in meat-type cockerels. An unselected, randomly bred population served as the control and provided tester females in all generations. By the eighth generation, males of the high line averaged at least four times as many completed matings as did low-line males; unselected controls were intermediate (McCollum et al., 1971).

Although mating ability is obviously a component of fitness, the results of Siegel's study indicate that natural selection prior to domestication had not exhausted genetic variation for this trait. As previously indicated, nature probably selects individuals with balanced phenotypes, rather than extreme types, thus preserving at least moderate levels of genetic variability, even for components of fitness. Although fertilization tests comparing the selected strains have apparently not been reported by Siegel's group, there is reason to wonder whether high mating frequency alone is a desirable criterion of selection for the breeder of livestock. Excessive mating activity may conceivably deplete the sperm supply of a male to such an extent that some matings would not result in fertilization.

Agonistic Behavior. As indicated earlier, agonistic behavior includes both aggressive and submissive or escape behavior. It is involved in establishment of social hierarchies or dominance orders which apparently exist within all or most group-living animals, including domesticated species. In situations where animals feed together or compete for necessities, it may significantly influence individual welfare and productivity. Sanctuary (1932) demonstrated the biological value of high social rank in flocks of hens; those in the upper half of the peck order laid more eggs than did those in the lower half. Similar associations have been found in varying degrees under a variety of conditions and in several species when competitive conditions exist. Such effects may be large when groups contain diverse behavioral types, as when different breeds are kept together in close confinement (Tindell and Craig, 1959).

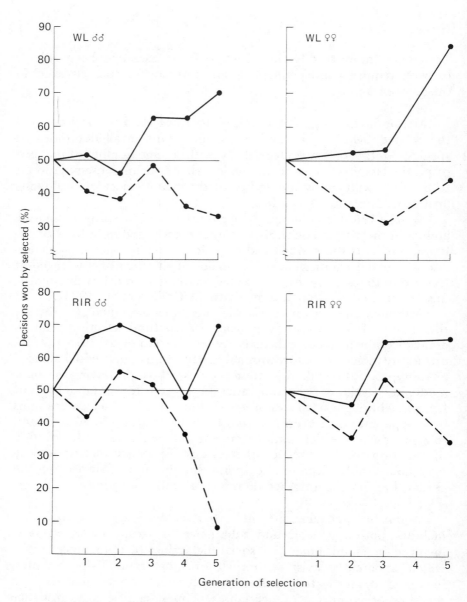

Figure 7-4 Percentage of pair contest decisions won by selected strain males (♂♂) and females (♀♀) of the White Leghorn (WL) and Rhode Island Red (RIR) breeds of chickens when matched with unselected controls. Selected strain performance over generations are shown as: Highs, solid lines; lows, dashed lines. The somewhat erratic results from generation to generation are probably caused by the relatively small numbers tested within strains and generations. However, the overall trend toward increasing differences over generations indicates that selection was effective. (From Craig et al., 1965.)

Selection for high and low social dominance within strains derived from three different foundation stocks of chickens produced large effects when continued for four or five generations (Guhl et al., 1960; Craig et al., 1965). Although realized heritabilities were relatively low, averaging about 0.20, high-strain individuals were dominant to lows in 75 to 100% of pair contests between those strains in the final generations.

Figure 7-4 illustrates results of the bidirectional selection study by Craig et al., which is fairly typical of studies where the trait under study has a relatively low heritability. Samples of each of the four selected strains were matched with unselected strain control birds of the same breed. Although differences between the high and low social dominance strains were somewhat erratic on a generation-by-generation basis, probably caused (at least partially) by the relatively small size of the samples tested, the overall results appeared to warrant these conclusions.

1. Five generations of selection for high and low social dominance ability produced large strain differences within each of two breeds.

2. Responses were essentially symmetrical; selection was effective for both increased and decreased social dominance.

3. The continuing response over five generations of selection, together with persistence of intrastrain variability, suggests polygenic inheritance.

CORRELATED RESPONSES

When strong selection pressure is consistently and successfully applied for one or a few closely related traits, other characteristics are likely to be changed also. Thus, long-term selection for running speed over moderately lengthy courses (as in Thoroughbred horses and in Greyhounds) has produced speedy runners and animals that have long legs and trim bodies. Selection for pulling strength and staying power in draft horses has resulted in moderate-length legs and large, muscular bodies. Those associations are readily understood, as form and function obviously go together in such cases.

In recent years it has become apparent that changes in some secondary traits (those not selected directly) may be undesirable, and that either a price must be paid, if certain extreme types are produced, or less extreme goals should be acceptable. As an example, extreme selection pressure has been applied by turkey breeders in recent decades for width of breast and rapidity of growth. Those traits have responded well to selection, but as turkey toms became broader-breasted and faster-growing, they also became relatively phlegmatic and awkward. As a result, male turkeys have become poorer breeders over succeeding generations of selection. Hatchery flock owners are now required to perform artificial insemination to obtain acceptable levels of fertility. Nevertheless, they willingly pay for the cost of artificial insemination, as narrow-breasted turkeys are no longer acceptable to the consumer, even if they do have superior sexual behavior and higher fertility!

Correlated responses, such as those described, are most likely caused by pleiotropic effects (i.e., particular alleles affect more than one trait).

GENOTYPE-ENVIRONMENT INTERACTIONS

A genotype-environment interaction occurs when strains or breeds perform differently, relative to each other, in different environments. Figure 7-5 illustrates three distinct situations involving pairs of genetic stocks as compared in pairs of environments. In Example 1 there is no interaction; both stocks do better in environment A than in B, but X and Y differ by the same amount in either environment. Example 2 illustrates a genotype-environment interaction. Stock W is very superior to Z in environment C and only slightly better in environment D. Stock W would be preferred in either environment, but it would not matter as much in D as in C. Example 3 represents the extreme case of an interaction; strain U is superior in environment E and V is better in F. In this situation we cannot say that one stock is better than the other without specifying the environment. Pfaffen-berger et al. (1976:174-175) point out that their German Shepherd dogs selected for guide-dog training would probably perform better for that purpose than would a strain selected for guard duty, but the reverse situation would probably exist if the same two strains were compared for usefulness as guard dogs.

Cook and Siegel (1974) compared three genetic stocks for mating ability of cockerels as adults, after sample halves of each stock were reared in either all-male or bisexual flocks. Their results are shown in Figure 7-6. The three lines had the same ranks when tested, regardless of method of rearing. As expected, the line selected for high mating frequency had the highest percentage of males mating once or more (in eight 10-minute trials) and the low-mating-frequency line had the

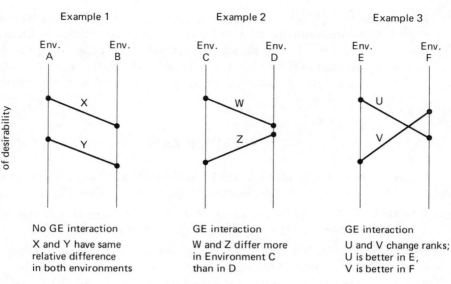

Example 1   Example 2   Example 3

Env.   Env.   Env.   Env.   Env.   Env.
 A   B   C   D   E   F

X and Y   W and Z   U and V

Phenotypic scale
of desirability

No GE interaction   GE interaction   GE interaction
X and Y have same   W and Z differ more   U and V change ranks;
relative difference   in Environment C   U is better in E,
in both environments   than in D   V is better in F

Figure 7-5 Models illustrating the presence or absence of genotype-environment (GE) interactions. Within each example, mean performances of each stock are shown on scales of phenotypic merit within each of two environments. GE interactions are present when stocks perform differently, relative to each other, in different environments (as in Examples 2 and 3).

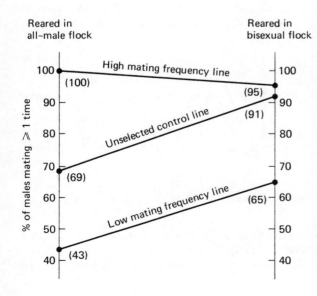

Reared in
all-male flock

Reared in
bisexual flock

High mating frequency line

Unselected control line

Low mating frequency line

(100)   (95)
        (91)
(69)
        (65)
(43)

% of males mating ⩾ 1 time

Figure 7-6 Effect of rearing environment on the percentage of males mating for each of three genetic stocks. A genotype – environment interaction is evident, as differences between stocks were greater when cockerels were reared in all-male flocks than when they were reared in the presence of pullets. (Drawn from data presented by Cook and Siegel, 1974.)

lowest percentage, regardless of how they were reared. Nevertheless, a genotype-environment interaction of the type shown under Example 2 of Figure 7-5 was present; a greater difference was present between the high and low lines when cockerels of those two stocks had been reared in all-male flocks than when samples of both lines were reared with females.

REFERENCES

COOK, W. T., and P. B. SIEGEL, 1974. Social variables and divergent selection for mating behavior of male chickens. Anim. Behav. 22:390-396.

CRAIG, J. V., L. L. ORTMAN, and A. M. GUHL, 1965. Genetic selection for social dominance ability in chickens. Anim. Behav. 13:114-131.

DALY, M., and M. WILSON, 1978. Sex, Evolution, and Behavior. Duxbury Press, North Scituate, Mass.

DICKSON, D. P., G. R. BARR, L. P. JOHNSON, and D. A. WIECKERT, 1970. Social dominance and temperament of Holstein cows. J. Dairy Sci. 53: 904-907.

FALCONER, D. S., 1960. Introduction to Quantitative Genetics. The Ronald Press Company, New York.

FULLER, J. L., and W. R. THOMPSON, 1960. Behavior Genetics. John Wiley & Sons, Inc., New York.

GODDARD, M. E., and R. G. BEILHARZ, 1974. A breeding programme for guide dogs. Proc., 1st World Congr. Genet. Appl. Livestock Prod., Madrid, pp. 371-376.

GUHL, A. M., J. V. CRAIG, and C. D. MUELLER, 1960. Selective breeding for aggressiveness in chickens. Poult. Sci. 39:970-980.

HODGSON, R. E., 1935. An eight generation experiment in inbreeding swine. J. Hered. 26:209-217.

McCOLLUM, R. E., P. B. SIEGEL, and H. P. VAN KREY, 1971. Responses to androgen in lines of chickens selected for mating behavior. Horm. Behav. 2:31-42.

PFAFFENBERGER, C. J., and J. P. SCOTT, 1959. The relationship between delayed socialization and trainability in guide dogs. J. Gen. Psychol. 95: 145-155.

——, J. L. FULLER, B. E. GINSBERG, and S. W. BIELFELT, 1976. Guide Dogs for the Blind: Their Selection, Development and Training. Elsevier Scientific Publishing Co., Agricultural Sciences Dept., Amsterdam, The Netherlands.

SANCTUARY, W. C., 1932. A study in avian behavior to determine the nature and persistency of the order of dominance in the domestic fowl and to relate these to certain physiological reactions. M.S. thesis, University of Massachusetts, Amherst, Mass.

SCOTT, J. P., and J. L. FULLER, 1965. Genetics and the Social Behavior of the Dog. University of Chicago Press, Chicago.

SIEGEL, P. B., 1965. Genetics of behavior: selection for mating ability in chickens. Genetics 52:1269-1272.

——, 1975. Behavioral Genetics. *In* E. S. E. Hafez, (Ed.), The Behaviour of Domestic Animals, 3rd ed. The Williams & Wilkins Company, Baltimore.

THOMAS, C. H., W. L. BLOW, C. C. COCKERHAM, and E. W. GLAZENER, 1958. The heritability of body weight, gain, feed consumption, and feed conversion in broilers. Poult. Sci. 37:862-869.

TINDELL, D., and J. V. CRAIG, 1959. Effects of social competition on laying house performance in the chicken. Poult. Sci. 38:95-105.

VAN VLECK, L. D., 1977. Principles of selection for quantitative traits. *In* J. W. Evans, A. Borton, H. F. Hintz, and L. D. Van Vleck (Eds.), The Horse. W. H. Freeman and Company, Publishers, San Francisco.

WILLHAM, R. L., D. F. COX, and G. G. KARAS, 1963. Genetic variation in a measure of avoidance learning in swine. J. Comp. Physiol. Psychol. 56:294-297.

8

Motivation

BIOLOGICAL NECESSITY AND RESPONSE

Behavior attempts to satisfy needs or drives. An animal with sufficient drive is prepared to "move"; it will respond to an adequate stimulus. However, a particular object or situation may be an adequate stimulus or not, depending on the strength of the drive in question. Thus, a particular food may be eaten or not depending on how long it has been since the animal's last meal, the palatability of the food, and perhaps whether it is even recognized as food. The amount eaten and rapidity of eating are associated with the level of motivation.

Drives commonly compete and when of equal strength may cause displacement activity, as discussed in Chapter 6. More commonly, coexisting drives are likely to be of unequal strength, and the behavior seen reflects the strongest. Life preservation and injury avoidance are high-priority drives, and instinctive reflexes commonly produce appropriate behavior when required. Thus, animals jerk away from painful objects, swim after falling into deep water, and seek food soon after birth. A strong hunger drive will usually take precedence over sexual activity. Also, the urge to explore strange surroundings is stronger than the need to play among young animals.

Stimuli cause behavior in motivated animals by either satisfying a need directly or because they have become associated with the satisfaction of a drive through learning or experience. Thus, hungry steers in a feedlot soon associate the coming of a feed truck with being fed. They may move to the feed trough when any truck approaches (generalization) or may learn to come for the feed truck only and ignore other vehicles (discrimination).

Regulatory Behavior

The internal environment of animals must be maintained within relatively narrow limits to ensure well-being. Homeostatic (steady-state) mechanisms have evolved to ensure relatively constant internal body temperature and the appropriate content of water, oxygen, and chemical constituents of cells, tissues, and body fluids. Monitor and control centers for such regulation are located in the hypothalamus and adjacent areas near the base of the brain and just above the pituitary gland. The pituitary is often called the "master" endocrine (hormone-secreting) gland of the body, and the hormones secreted by it usually regulate other endocrine glands. It is now clear that the pituitary is controlled, in turn, by the hypothalamus.

Regulation of vital body functions occurs at two major levels. Sensory and nerve cells convey information from all parts of the body to the brain. When significant deviations from the optimum are detected, the autonomic nervous system and the pituitary gland receive signals from the hypothalamic or nearby regions which cause physiological responses. If those responses are not adequate to bring the body into equilibrium, then the second level of regulation, consisting of behavioral responses, is activated. Thus, in an animal subjected to cold, blood will be shunted away from the body surface and involuntary shivering may occur (this muscular activity increases heat production). If the body temperature continues to fall, the animal will behave in a manner likely to increase its warmth, such as moving to a less exposed location, into a group for bodily contact and insulation, or by otherwise altering its environment (Figure 8-1).

Environmental Priming

Changes in day length, nutritional status, and average temperature can have large indirect effects on behavior, particularly for species or breeds that developed in environments where seasonal changes were large. Under such conditions, the animal's physiology and general sensitivity to certain stimuli are altered. Behavioral changes are often mediated by hormonal responses that alter the animal's neuro-muscular sensitivity and set physiological processes in motion (or cause them to slow or cease). Reproductive processes, in particular, are likely to be modified. Hormonal effects are considered in later chapters.

Seasonal changes in day length affect reproductive behavior in sheep, goats, and horses in ways summarized by Fraser (1968:40-46). Those breeds of sheep and goats developed in the temperate zones

103

Figure 8-1 (a) A cold pig turns on heat lamps by inserting its nose in the hole in the apparatus. (b) Penmates bask under heat lamps after one has turned on the heat. (Courtesy of S. E. Curtis, University of Illinois.)

show maximum sexual activity in late summer or autumn. That breeds may be rather specifically adapted to particular environments was demonstrated by about 8 weeks' earlier onset of sexual receptivity of Suffolk than of Cheviot ewes when samples of both breeds were kept at the same northerly location of Edinburgh, Scotland (Figure 8-2). The Suffolks, developed in southern England, exhibited average onset of estrus in August and would, if mated at first heat, lamb in midwinter, whereas the Cheviots would lamb in early spring. The Cheviots are better adapted to Scottish conditions, as environmental priming of their reproductive physiology and associated sexual behavior requires shorter day lengths (or a longer interval of short days).

It has been shown that sheep can be brought into estrus outside their natural breeding season by artificial reduction in daily light,

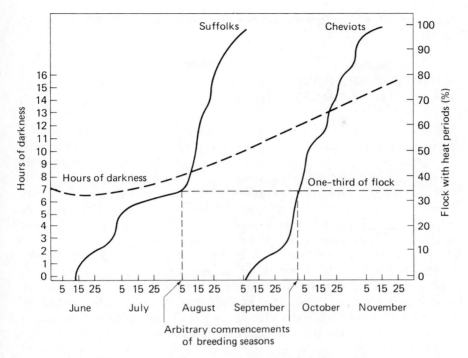

Figure 8-2 Rates of commencement of heat periods throughout two flocks of different breeds at the same geographical location (Edinburgh) in the same year of study. The period when one-third of the flock has shown heat approximately coincides with the date assumed by local commercial breeders to be the commencement of the true breeding seasons of the breeds concerned. [With permission from *Reproductive Behaviour in Ungulates*, A. F. Fraser, 1968. Copyright by Academic Press, Inc. (London), Ltd.]

but 3 to 4 months may be required to obtain a response. Horses also respond reproductively to changes in amount of light received daily, but in opposite fashion to sheep; as long-day breeders, they require increased light (either natural or artificial) to induce estrus and reproductive functioning.

Poultry are greatly influenced by day length in age at attainment of sexual maturity and in persistency of egg production. Thus, the onset of egg production can be modified by at least a month by rearing chicks under short- or long-day artificial lighting, and decreasing light is likely to cause the onset of moulting (loss and replacement of feathers) in hens that have been laying for some months.

REWARD, PUNISHMENT, AND MOTIVATION CENTERS

Study of the brain's control over behavior may involve removal or damage of particular regions to see what responses are lost. Alternatively, stimulation of those same regions can be done by electrical, hormonal, or other means to determine what behaviors are elicited. Other methods, such as recording of brain waves during different kinds of activity and during sleep, are also useful. Physiological psychologists involved in such studies face challenging problems and are only beginning to unravel the brain's mysteries.

"Reward centers" were accidentally discovered by stimulating subcortical areas of the brain (Olds and Milner, 1954). It was observed that a rat would persistently return to the area of the cage where it had first been stimulated. That brain stimulation could be rewarding was clearly demonstrated by placing a rat with an electrode implanted in its hypothalamus in an apparatus where it could activate current to the electrode by pressing a bar. One animal responded by pressing the bar at an average rate of 2000 per hour for a 24-hour period; it then slept most of the next day (Olds, 1958). Subsequent to the work of Olds and Milner, many anatomical regions have been explored to find those eliciting self-stimulation. In addition to numerous pleasure centers, punishment centers have also been located. Milner (1970: 378–412) presents a review of the general topic and states: "Rats will learn mazes, make discriminations, cross electrified grids, and press levers for brain stimulation reward."

Reward centers may also represent motivation areas, as indicated by the ability of the experimenter to cause eating, drinking, copulation, and other specific behaviors by stimulating particular subcortical brain areas. When the means of satisfying a drive (such as hunger) is not available, the stimulated animal may become active (e.g., it may run in an exercise wheel, as if showing appetitive behavior).

INNATE RESPONSES

Why do certain stimuli, not previously encountered, cause appropriate behaviors? In general, nature motivates naive animals into giving correct responses when stimulated for situations where learning how to respond would be too hazardous or too slow to assure fitness. Life-styles enter the picture in important ways. In some species, relatively low on the evolutionary scale, the young must survive without parental care; continuity of the species in such cases depends on instinctive behaviors. Understanding the immediate causes of many such behaviors is in its early stages, but it is known that subcortical brain areas, involving the hypothalamus in particular, influence physiological regulation and how animals are aroused by stimuli acting on motivation centers to various kinds of activity. Although learned responses allow "fine tuning" of an animal's behavior and are often important in domestic animals, we should be aware that many essential responses occur initially as innate behavior.

Characteristically, innate behavior is adaptive, stereotyped within species (sometimes within one sex) and may be exhibited at particular ages or in certain physiological states when stimulated by "releasers" or "unconditioned stimuli." Some familiar instinctive behaviors and their primary functions are listed in Table 8-1.

Table 8-1   *Some familiar instinctive behaviors in domestic animals and their primary functions*

Animal	Instinctive behavior	Functions
Newly hatched chickens and turkeys	Pecking at and swallowing small, bright objects	Location and ingestion of food
	Pecking at reflective surface	Location of water
Newborn mammals	Searching movements of head and neck, sucking on protruding objects, swallowing fluids	Location and feeding on nipples of mammary gland
Young birds and mammals able to see and control their movements	Avoidance of cliffs and holes	Prevention of falling and injury
Newborn and very young pups	Urination and defecation when rubbed by a warm, wet object	Elimination of body wastes at appropriate time
Mother dog	Licking of anogenital regions of newborn and young pups and consumption of wastes	Causes elimination by young and avoids soiling of the "nest"
Sexually mature male dog	Lifting of hind leg and urination on conspicuous objects	Communication of presence of adult male in the area
Sexually mature gilt	"Standing" for mating when reproductively appropriate in the presence of sexually mature boar	Stimulation of sexual mounting and ejaculation by boar when eggs are available for fertilization

Duration of Instinctive Behavior

Many innate behaviors persist for typical periods of time. Thus, suckling, once initiated in young mammals, continues for a fairly definite period characteristic of the species involved. Early weaning and feeding by methods other than the use of artificial nipples frustrates the suckling response. Baby pigs, calves, pups, and human beings (among others) typically suck on any readily available substitute when they are weaned at an early age. Parts of the bodies of other young or of the animal's own body may be sucked preferentially, probably because the warmth and softness resemble that of the mammary gland. Calves fed from buckets will often develop the habit of sucking on other calves if they are present, at times with adverse effects on their health and well-being (balls of hair and other inappropriate items may be found in the stomach). In the absence of other animals, they may suck on steel chains, iron bars, or any other object that can be taken into the mouth. Eibl-Eibesfeldt (1970: 48–49) reviews observations on the suckling behavior of human babies. When fed by nipples attached to bottles of milk, infants having adequate intakes after 20 minutes were satisfied and slept. However, when nipple openings were larger, allowing adequate intake in 5 minutes, the babies remained unsatisfied and continued sucking movements after the bottle was removed. Given an empty bottle, they would suck for another 10 to 15 minutes and then appeared to be satisfied. (It appears that in human societies in which breast feeding is acceptable, thumb sucking of babies is not a problem.)

Maturation

Many behaviors, initially instinctive, show improvement in efficiency or are altered by experience and learning. Thus, the accuracy of pecking and choice of items eaten by chicks and poults improve rapidly with age and with learning which items are food.

Some behaviors cannot be expressed until an appropriate developmental stage has been reached. Differences between precocial and altricial animals are clearly evident here. Neonatal kids and day-old chicks usually avoid falling over edges, but Scott (1958) observed that puppies placed on scales may crawl to the edge and fall off time and again when in the neonatal and transition stages of development. During these stages, from birth to about 3 weeks, puppies are either blind (eyes open at about 2 weeks) or see poorly and have difficulty in learning.

Sexual maturity and hormonal priming may also be required.

Among dogs, castrated individuals and bitches do not ordinarily show the leg lifting and scent marking of posts and other conspicuous objects that is so common among sexually mature males. Signoret (1970) has shown that ovariectomized gilts reared in isolation and then injected with estrogen were attracted to boars and showed immobility reactions associated with copulatory behavior comparable to intact and socially reared females in estrus. However, Signoret points out that early deprivation of social contact impairs the sexual behavior of male primates, rats, and guinea pigs. Apparently, the male, having a more complicated and demanding role in sexual behavior, must rely on both instinctive and learned elements in achieving satisfactory copulation.

Releasers and the Innate Releasing Mechanism

Ethologists have dealt extensively with the concepts of innate releasers (also called key stimuli or sign stimuli), innate releasing mechanisms, and instinctive responses. Psychologists interested in learning theory use related concepts but different terms; they speak of unconditioned stimuli, neural (or hormonal) pathways, and unconditioned responses. We may compare the usages as shown in Figure 8-3.

The *innate releasing mechanism* presumably exists within the brain and allows instinctive behavior to occur when acted upon by an appropriate releaser. In the examples given in Table 8-1, some of the releasers are: small, bright objects and reflecting surfaces (for pecking and ingestive behavior of chicks and poults); courting chants;

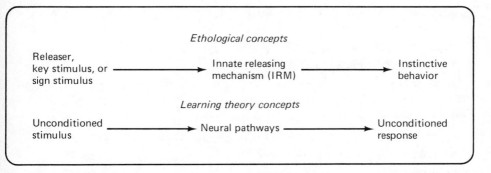

Figure 8-3 The expression of unlearned behaviors that fulfill essential requirements are viewed in related ways by ethologists and psychologists, but different terminology is used, as indicated.

boar odor; and tactile stimulation or any combination of boar-produced stimuli (for immobility stance of estrous gilts). By analogy, the motivated animal may be compared to a cocked and loaded gun. When the releaser acts (pressure is exerted), the innate releasing mechanism responds (trigger is pulled), and instinctive behavior occurs (the gun fires).

Supernormal releasers are exaggerated stimuli that are more effective than normal or natural releasers. Gaping mouths release feeding behavior in altricial birds caring for their young. The hungrier the baby bird, the wider the mouth is opened (unless weakness from prolonged hunger occurs). Parasitic birds such as cuckoos and cowbirds have young that show extremely large, gaping mouths when hungry and are therefore preferentially fed by their unknowing foster parents. A supernormal releaser of practical importance is commonly used in the collection of bull semen. Artificial vaginas are heated to about 115°F, thereby causing bulls to ejaculate more readily than if they had temperatures comparable to a cow's vagina (from 101 to 102°F).

REFERENCES

EIBL-EIBESFELDT, I., 1970. Ethology: The Biology of Behavior. Holt, Rinehart and Winston, New York.

FRASER, A. F., 1968. Reproductive Behaviour in Ungulates. Academic Press, Inc., London.

McCLEARY, R. A., and R. Y. MOORE, 1965. Subcortical Mechanisms of Behavior. Basic Books, Inc., New York.

MILNER, P. M., 1970. Physiological Psychology. Holt, Rinehart and Winston, New York.

OLDS, J., 1958. Satiation effects in self-stimulation of the brain. J. Comp. Physiol. Psychol. 51: 320-324.

——, and P. MILNER, 1954. Positive reinforcement produced by electrical stimulation of septal area and other regions of rat brain. J. Comp. Physiol. Psychol. 47: 419-427.

SCOTT, J. P., 1958. Critical periods in the development of social behavior in puppies. Psychosom. Med. 20: 42-58.

SIGNORET, J. P., 1970. Sexual behaviour patterns in female domestic pigs (*Sus scrofa* L.) reared in isolation from males. Anim. Behav. 18:165-168.

9

Socialization

Social behavior begins before hatching in some birds. Fetuses of quail, chickens, ducks, and geese respond to auditory or vibrational signals. As they approach hatching, they begin to make peeping and "clicking noises" (associated with breathing) within the shell. The clicking noises, in particular, are effective in synchronizing hatching of the baby birds from eggs that are in contact. Vince (1973) found that when pairs of eggs are together, the young hatch within a closer interval of time than if they are separated. When its own egg is in contact with another given 24 hours more incubation, the fetus in the "stimulated" egg hatches about 10 hours earlier, on the average, than it would otherwise. Artificial clicks, administered at the rate of three per second to isolated eggs late in the incubation period, are even more effective (a supernormal releaser), advancing hatching time by more than a day. Gottlieb (1965), using tape recordings of maternal hatching calls, has demonstrated that unhatched ducklings have increased rates of peeping and otherwise respond in accelerated ways when stimulated by those recorded sounds.

Synchronized hatching is highly adaptive for precocial birds; those hatching very late are left behind when the dam leads her brood away from the nest. Precocial birds and mammals are already well advanced at hatching or birth and are able, within a few hours or days, to follow their own kind. The following response, called imprinting, is an expression of early socialization or primary socialization. Although most domestic animals are precocial at birth, the concept of "critical periods" for socialization will be considered with examples based on the development of such behavior in dogs, which are less developed at birth and go through the process in "slow motion" as compared to precocial species.

CRITICAL PERIODS

The theory of critical periods in biology and psychology was developed near the beginning of the twentieth century. It will be considered only briefly here as set forth by Scott and his associates. The theory is based primarily on observation and in its simplest form (Scott et al., 1974) states that "the organization of a system is most easily modified during the time in development when organization is proceeding most rapidly." A likely situation is that socialization occurs at equivalent developmental ages of precocial and altricial species, although those occur at different chronological times relative to birth, as shown in Figure 9-1. The critical period for primary socialization occurs during that interval when animals are first able to control their movements with some skill, can make full use of their senses (especially sight and hearing), and can learn to identify and remember other individuals. Because of the lack of precise limits, some prefer the terminology "sensitive periods" rather than critical periods. Secondary socialization occurs later, but differs qualitatively and develops at a slower rate.

In studying the socialization of wolves to man, Woolpy and Ginsburg (1967) found that wild-caught adult wolves can be socialized, although the process required several weeks. The adult wolves then behaved toward human beings as if they were interesting and rewarding objects. On the other hand, wolf cubs, taken from the litter before their eyes opened (or soon after) and raised by hand, socialized rapidly as soon as the critical period began and behaved toward man as they would toward dominant members of their natural wolf pack; they developed what might be called a filial attachment. As indicated above, socialization during the early stage is called *primary* and occurs during the critical or sensitive period; socialization at later ages is called *secondary*.

That permanent socialization may require more than a brief exposure during the critical period was indicated in the Woolpy and Ginsburg study. Young cubs that readily socialized with people during the critical period were returned to the pack and later behaved the same as wild wolves when adults.

Onset of the Critical Period

Puppies do not object to being separated from their litter and rearing place at 2 weeks, but will bark or whine for prolonged periods if separated from either at 3 weeks of age or later. This separation

(a) *Precocial species* such as chicken, turkey, horse, sheep, cattle.

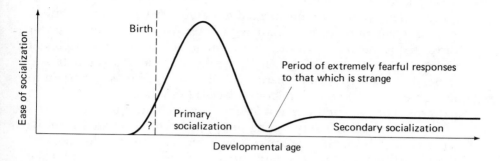

(b) *Altricial species* such as dog, cat, rabbit.

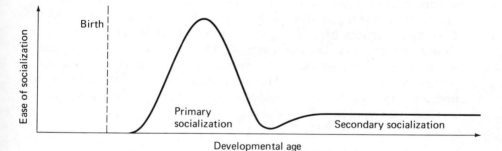

Figure 9-1 Effect of developmental age on ease of socialization. Although occurring at different times relative to birth, the period of primary socialization begins at about the same developmental age in precocial and altricial species, when the young are first able to see and hear well, can control their movements with some skill, and are able to identify and remember others. The period of primary socialization ends as the young animal becomes fearful of that which is strange and is followed by the period of secondary socialization, which differs in ease of establishment of social attachments and in qualitative ways from primary socialization.

distress indicates that strong attachments have already occurred by 3 weeks. After 2 weeks, puppies have their eyes open and by 3 weeks give startle responses to unexpected noises, indicating hearing ability.

Although puppies can learn with reference to sucking behavior early in the neonatal period, their general ability to learn begins at about 18 or 19 days of age. By 3 weeks they can discriminate and remember what they have learned. Brain-wave patterns also begin to resemble those of adults at about 3 weeks (although the transition to patterns typical of adults is not attained until about 8 weeks).

In summary, the very young puppy is largely insulated from its environment by immature sense organs, incomplete development of the central nervous system, and the protective care provided by the mother. By 3 weeks of age it has developed sufficiently so that it can learn in much the same way as adults and is ready to form strong social attachments; the critical period for primary socialization has already begun (Scott et al., 1974).

Precocial mammals have their senses relatively well developed at birth, and precocial birds, as noted above, are able to respond behaviorally to sounds emitted by other members of their species even before hatching. The critical period for socialization in precocial species begins either before or shortly after birth, as compared to the later beginning of the period in altricial animals (Figure 9-1). The phenomenon of following, as soon as muscular coordination allows, particularly the following of objects emitting sounds characteristic of the species, is an indicator of the early socialization process in precocial animals. However, birds can form attachments to immobile as well as moving objects, indicating that following is not an essential requirement for the process.

End of the Critical Period

How can the end of the critical period for primary socialization be determined? Freedman et al. (1961) placed pregnant mother dogs in fields with high board fences and allowed them to whelp and rear their puppies without direct human contact, except when one at a time was removed at certain ages (2, 3, 5, 7, and 9 weeks) and given the opportunity to establish social bonds with man during the following week. Pups were then returned to their own litters until 14 weeks, when all were tested individually. Puppies removed at 2 and 3 weeks were relatively unresponsive, but those taken out at 5 and 7 weeks socialized readily and were still responsive to man at 14 weeks. Those removed at 9 weeks did not socialize as readily then or at 14 weeks. Control pups, not exposed to man before 14 weeks, were extremely fearful and never formed close attachments to human beings.

From the study cited above and others, it is now believed that the best time for socialization of a puppy to another species is between 6 and 8 weeks and that primary socialization requires more time and effort at older ages. The critical period comes to an end at about 12 weeks, on the average. Breeds and individuals may differ in their behavioral development by a week or more, and the period may be extended under special circumstances.

What brings the critical period for socialization to an end? Scott et al. (1974) believe that dogs can be socialized at any age after 3 weeks if interfering factors can be reduced or eliminated. Four factors are recognized by them as follows:

1. *Fear:* The young puppy becomes increasingly fearful of that which is strange; this is particularly noticeable beginning at about 7 weeks and reaching a maximum at about 14 weeks.

2. *Separation distress:* After social attachments (and attachments to particular places) have become established, the animal becomes emotionally upset when separated.

3. *Previous attachments:* If not removed from the presence of other animals or places to which attachments have formed, the animal directs its attention to those, thereby largely excluding contacts with new individuals (or places).

4. *Defense:* With the development of sexual maturity, the animal may become very defensive of its "own" territory, as shown by a tendency to attack or threaten strangers.

The first three interfering factors are probably of importance in most other domesticated species.

THE PRIMARY SOCIALIZATION PROCESS

Do young animals form their initial social bonds because of food rewards? In mammals the young nurse and in birds the young are either fed by the parents (altricial species) or are led to food by the mother (precocial species). That food rewards are not required for primary socialization has been shown in numerous studies with mammals, and the following response generally begins in precocial birds before they feed. Scott (1962) reviewed several studies and concluded that food rewards *may* contribute to the formation of social relationships, although they are not of prime importance.

Apparently all that is required for primary social attachment to occur is that the animal must have the use of its senses so that it can discriminate between familiar and unfamiliar objects, and that it must be able to remember those that are familiar (Scott et al., 1974). Learning is clearly part of the process, but what drives are involved? As will be seen in Chapter 10, either positive reinforcers (rewards) or negative reinforcers (punishment) are usually required for associative learning to occur. But the very young animal presents a puzzle in this regard. Puppies learn to run and show improvement when the goal is a completely passive individual (Stanley and Elliot, 1962); thus the reward may simply be in having "company."

A surprising finding, from studies involving several species of birds and mammals, is that the process of primary socialization is not usually inhibited by punishment (Scott, 1962). In fact, any strong emotion experienced by the young animal, such as fear, pain, loneliness, or hunger, usually speeds up the process. Young chicks following a motorized model will approach even closer if the model emits loud sounds or if a mild electric shock is given while the chick is following.

Reversibility of Imprinting

Although some characteristics of attachment behavior were described by earlier investigators, Lorenz (1937) established himself as an authority in this area of research and stimulated many additional studies. Among his fascinating observations, it is revealed that he had a pair of Graylag geese that were induced to hatch eggs laid by a Muscovy duck. The ducklings associated with the geese until about 7 weeks of age, but gradually lost interest in their foster parents and socialized thereafter among themselves and with other Muscovies in ordinary duck fashion. However, at 10 months, the only male in the group showed sexual behavior toward and attempted to copulate with Graylag geese (whether male or female) rather than with Muscovy females. Lorenz regarded the "imprinting" response as irreversible and stated that "it is the chief biological task of imprinting to establish a sort of consciousness of species in the young bird, if we may use the term 'consciousness' in so broad a sense." He proposed that during a very brief period, early in the life of birds, long-lasting filial and sexual attachments become established as part of the same process.

Later studies indicate that filial and sexual attachments may occur at different times; sexual attachment often occurs later and may require a longer exposure (Immelmann, 1972). Thus, for Graylag geese the critical period for sexual attachments begins at about 50 and continues to about 140 days, whereas in Mallard ducks it begins between 9 and 19 days and continues for several weeks.

Sexual preference for the "wrong" species may be present in birds because of early exposure to that species, yet matings may occur with the bird's own species in the absence of distractions. Lorenz noted a few cases in birds in which an individual's sexual reactions, although directed primarily toward man (because of hand rearing), would finally be directed toward a member of its own species as "a rather poor substitute for the beloved human." This was further demonstrated by Schein (1963), who studied male turkeys reared apart from other turkeys, but with considerable human contact, for the first 32 days after hatching. After 5 years, the three turkey toms involved would court humans rather than turkey hens when given the choice. When the human observer was not present, the toms would then court turkey hens.

Mammalian filial and sexual attachment behavior appears to be more easily altered. Sheep have long been recognized as readily forming social bonds with man or other species. Cairns and Johnson (1965) separated some 4- to 8-week-old lambs from their dams and placed them with dogs (lamb–dog pairs were isolated). After 9 weeks of cohabitation, lambs kept with dogs were compared with lambs that had been kept with ewes (as pairs). When given a choice between previous cohabitants and an animal of the other species, lambs chose their familiar companion in nearly all cases, whether it happened to be a dog or a ewe. Cairns (1966) presents a review of studies dealing with attachment behavior in mammals.

MOTHER-YOUNG BONDING

A typical feature of maternal behavior in domesticated animals is that the prospective mother, if allowed to follow her natural inclination, withdraws from her ordinary social group before the young are hatched or born. Broody hens are naturally isolated on their nests because of incubation requirements. Bitches, cats, and sows find a favorable, secluded spot and construct a "nest" (i.e., some kind of depression to hold the young). Ewes and cows will separate themselves from the group, but mares may or may not. The latter species make only minimal efforts to find cover. However, in nearly all cases, separation from the group is achieved, allowing mother and young at least a brief period of peaceful isolation during which social bonding may occur between them.

Feral hens and their broods were observed by McBride et al. (1969). Broods were generally well spaced so that there was little chance of mixing. When drawn together at sites where feed was provided, hens usually prevented contact between groups by charging strange chicks. In one instance, when two broody hens came together, one hen chased the other. The young chicks of the two hens intermingled during the incident with apparent indifference. Following their interaction, the two hens returned and gathered their own chicks together by tidbitting. Thus, mother–young groups of chickens are generally maintained separately by the behavior of the dams in avoiding contact. Clearly, hens recognized their own young, and chicks learn to identify their mothers, as indicated by attraction to their own mother hen after groups become mixed.

A common maternal behavior of most domesticated mammals is the consumption of birth fluids and vigorous licking of newborn. Although other important functions are served also, it seems that this contact provides olfactory and perhaps other cues that are important in identifying the dam's progeny (Figure 9-2).

Several studies on goats strongly suggest that the first 5 to 10 minutes after birth is a critical period for does in establishing attachment to their own young (e.g., see Klopfer et al., 1964). Most does, if allowed this brief postpartum period of intimacy with their kids, will then recognize and accept them following later separation for intervals up to 8 hours. However, when kids were removed immediately after birth and then returned after a few hours, many were actively rejected by their own mothers.

The critical period for "maternal imprinting" in sheep appears to last up to 8 hours after birth of the lamb. Smith et al. (1966) found that ewes will accept the first lamb presented to them after parturition (their own lamb or another ewe's) and that, if allowed to lick the wet lamb for 20 to 30 minutes, they will form a lasting attachment. Unlike goats, "separation of the lamb from the ewe immediately after birth for periods up to eight hours did not prejudice the subsequent physical development of the lamb." Those lambs that were removed experimentally and not cleaned by man were licked and nuzzled when placed with a ewe, which had been isolated from other sheep in the interim. (Lambs that were separated sucked on inappropriate objects before being returned to a ewe and often had difficulty in locating the mammary gland and nipple.)

The newborn ungulate, although easily socialized, is not very selective in its choice of a mother figure, as we have already seen. It

Figure 9-2 A cow separated from the herd during parturition, cleans her recently born calf (above) and then allows it to nurse alongside (below). The head-to-tail alignment of the mother and her very young offspring is characteristic of most ungulates. It allows positive identification of the young animal by its odor. Failure by the female to identify the young animal as her own usually leads to prompt rejection of its advances. (Photographs by the author.)

is likely that the temporary withdrawal of the dam from her social group serves at least three functions: (1) the dam has the opportunity to form a strong social bond with her offspring while licking and nuzzling it without disturbance by other animals; (2) other expectant mothers are prevented from "stealing" her young; and (3) the newborn animal has some opportunity to learn to identify its own mother. After having been thoroughly cleaned by its mother, the newborn animal becomes far less attractive to other females about to give birth, and the dam and her progeny may then safely return to the group with little likelihood of the young animal being led astray by mothers-to-be. Kilgour (1972) aptly describes these relationships among sheep and points out the hazard of mismothering, which may become acute under some recently developed management schemes: "When ewes are synchronized to lamb within a short period of time and are highly productive, giving a large number of multiple births, and when paddocks are small and all cover over three ft. high has been cut out, then possibilities for lamb poaching are very high. A study using Corriedales in Victoria under lowland conditions showed 15 percent of lambs to be swapped or mixed, with several deaths due to this cause."

Heifers and cows on pasture typically calve away from the herd; at times it appears as if they simply remain behind to give birth, as the herd moves along. After being licked clean and nursed, calves remain quietly in the location where they were born while their mother moves away to feed and drink (Figure 9-3). The dam may rejoin the herd, except for brief excursions to the birth site to care for the calf. After a few days the cow will then lead her calf back to join the herd. Although it may attempt at first to suckle from other lactating females, the calf soon learns to suckle from its own mother, after being butted or kicked away by others. Some of the same risks of mismothering, confusion in mother-young bonding, and resultant abandonment or loss of attachment are likely with synchronized calving under high-density conditions as were described for sheep under similar conditions by Kilgour (see pp. 119-120).

Observations of semiwild New Forest ponies in England by Tyler (1972) provide some insights into the mother-young bonding of horses under relatively natural conditions. Although some mares approaching parturition moved away from their companions, others foaled within a few yards of other ponies (which paid little attention to the birth process). Mares began to lick their foals within a few minutes after birth and continued for up to half an hour. The fetal fluids probably help to establish the dam's recognition of and social bonding to her offspring; after licking her foal, a mare could easily discriminate between her own and others (which were threatened whenever they approached). Although several mares did not separate themselves from the herd during foaling, they took care during the first day to stay close to their foals. Subordinate ponies that ap-

Figure 9-3 A very young calf remains quietly in seclusion while its mother grazes elsewhere or rejoins the herd. The calf is allowed to nurse a few times daily when its mother comes to visit and is led to the herd when a few days old. (Photograph by the author.)

proached were threatened, and the mare would lead her foal away if approached by a dominant mare; the mother would interpose herself between her foal and other animals (whether dominant or subordinate to her) during the first day. Tyler observed that after the first few suckling bouts, mares would often move away as their foals approached to suck but would finally remain still when the foal persisted in following. She speculated that "this probably aided the development of the mother–foal relationship by encouraging the foal to follow its mother; milk then acted as a reward for this behavior." It should be kept in mind, however, that experimental studies in other species indicate that development of primary socialization in the newborn does not depend heavily on food rewards, although they may have a minor effect.

Sows largely ignore their young while delivering the litter; baby pigs move to the udder unassisted and begin to nurse from the nipples within the first half-hour after birth. Although sows may defend their litter and respond to a pig's squeals, they apparently do not recognize their own progeny for the first couple of days and will readily accept any baby pig that comes their way during this early period. Perhaps the lack of licking and cleaning of the baby pigs may be involved? Arguing against this hypothesis is the observation that most bitches and cats will also accept the young of other litters for a few days after birth, although the dam vigorously licks and cleans her progeny soon after birth in those species.

ADOPTIONS

Occasionally, the need to foster the young of one mother onto another arises, as when the mother dies or is unable to adequately care for her own offspring. A number of recent developments and discoveries have increased the feasibility and desirability of including adoptions, particularly in sheep and cattle, as a part of some management schemes. Thus, Finnish Landrace sheep, noted for their great prolificacy, are being crossed with other breeds to increase the number of lambs born per ewe. When three or more are produced, it appears desirable to even out lambs per ewe by fostering. Many beef cattle producers are now breeding their heifers to calve at 2 years of age, thereby increasing lifetime calf production. In some cases 2-year-old heifers are greatly delayed in returning to reproductive cycling because of the demand of their calves for milk; fostering the calves of 2-year-olds onto more mature cows would allow more rapid recovery of cycling. The use of hormones for synchronizing estrous cycles and parturition in herds and flocks is increasing (e.g., see Beardsley et al., 1976); synchronized parturition increases the feasibility of fostering young, as it is most easily and effectively done when a number of dams are giving birth at or near the same time.

Forced Fostering versus Adoption

Sheep. Shepherds have known for a long time that orphan lambs covered with the pelt of a lamb that died are likely to be accepted by the mother of the deceased. That the sense of smell is of paramount importance in accepting or rejecting lambs at close quarters was shown by Morgan et al. (1974), who surgically deprived ewes of their sense of smell. They stated that "the ewe relies so heavily on the sense of smell to discriminate against alien lambs that without the use of this sense she will invariably accept any alien irrespective of size and in fact will accept two aliens simultaneously."

Neathery (1971), in a small but carefully controlled study, found that five of six Hampshire ewes adopted orphan lambs 2 days after giving birth when given a single, adequate dose of tranquilizer. Only one of six control ewes (not receiving the tranquilizer) adopted an orphan under otherwise similar conditions. Forced fostering, involving the confinement of a ewe with another's lamb in a pen, usually requires tying the ewe to prevent her from butting and abusing the lamb. This practice required an average of 10 days before acceptance in one study, although some ewes persisted in butting the alien lamb for over a month (Hersher et al., 1963). As indicated earlier, Smith's group found that the first "wet" lamb or lambs presented within an interval of 8 hours would be permanently accepted by a ewe.

Cattle. The effects of rearing two calves (one alien to the cow) as compared to one (the cow's own) for beef production were studied by Wyatt et al. (1977) using 6-year-old Hereford X Holstein cows. Cows rearing two calves were forced to foster the second calf, which was placed with the cow at the time of birth of her own. Cows rearing two calves yielded nearly 40% more milk and weaned 60% more total calf weight than those raising singles. The fostered calves suckled less frequently and for shorter periods and weighed less when weaned than the cows' own calves. Cross nursing was frequently observed among the cows that had been forced to accept a second calf, but this was not seen for cows rearing single calves. Cows raising two calves required more supplemental feed during the winter and had longer postpartum anestrus intervals.

Using information gained from previous studies on the development of maternal attachment, Hudson (1977) compared two methods of fostering four calves on each of 10 dairy cows. Six cows had their own calves removed at birth and the amniotic fluid was collected after piercing the second "water bag" as it appeared. The four calves to be adopted (up to 10 days old) were then wiped with the fetal fluid and placed with the cow. Although all cows showed some initial interest, three kicked and butted the calves vigorously. Nevertheless, all calves smeared with the amniotic fluid were being licked and nursed by the following morning, and the cows showed signs of distress when the calves were removed for weighing. In the second group, calves were force-fostered onto each of four cows by a more traditional method. The cow's own calf was removed within 3 hours of birth (presumably after having been licked and accepted). Four hungry calves were then introduced and the cow was restrained until all had suckled. Cows were subsequently restrained twice a day while calves suckled until each cow permitted all calves to suck without being restrained while confined in a small pasture. The individual cow–calf groups were then kept isolated an additional 10 days before being placed with the multiple-suckling herd.

Calves adopted within 24 hours because of removal of the cow's own calf and smearing with amniotic fluid benefited from the "motherly" behavior of the foster dam, which bonded to them as if they were her own calves. Traditionally fostered calves were accepted to varying degrees and there appeared to be considerable confusion as to proper pairings of foster dams and calves after they were placed in a multiple-cow herd on pasture. Table 9-1 presents some of the results. An average of three calves per cow was raised to 12 weeks in the traditionally fostered group, but all four calves survived in the bonded group. Suckling behavior in the traditional group appeared to be somewhat disorganized; there was a high incidence (30%) of suckling on cows other than the one to which fostering was attempted. Observations indicated that the bonded group calves fed simultaneously, so that each should have received an equal amount of milk. Cows of the group that formed specific maternal bonds with each of their four adopted calves weaned 332 kg of calves at 4 weeks. Those cows that were forced to foster calves weaned 233 kg of calves after the same interval.

Although Hudson's study was carried out with relatively few animals, it strongly suggests that the fostering of young animals can be done with a high level of success if those fostered are handled in such a way as to cause maternal bonding very soon after birth of the dam's own young and before she has formed such a bond with her

Table 9-1 *Behavior, mortality, and body weights of calves fostered by a traditional method as compared with those of foster calves where
maternal bonding was induced*

	Method of fostering	
Item	Traditional	Maternal bonding
Cows/treatment	4	5[a]
Calves fostered/cow	4	4
Calves reared to 12 weeks/cow	3[b]	4
Cross suckling (%)	30	3
Mean calf weight at 12 weeks (kg)	77.5	82.8
Total calf weight at 12 weeks (kg)	233	332

[a]One cow, which died of "grass staggers" 5 days after calving, is not included.

[b]Many calves in this treatment group scoured badly; four calves were lost or removed
within 8 days after fostering.

SOURCE: Data from Hudson (1977).

own offspring. Although investment of time is required during the
critical period just after birth, it may be worthwhile, as compared to
the considerable amount of time also required with traditional fostering methods and because of the superior performance of young animals that are accepted by the dam as her own.

REFERENCES

BEARDSLEY, G. L., L. D. MULLER, H. A. GARVERICK, F. C. LUDENS, and
W. L. TUCKER, 1976. Initiation of parturition in dairy cows with dexamethasone. II. Response to dexamethasone in combination with estradiol
benzoate. J. Dairy Sci. 59:241-247.

CAIRNS, R. B., 1966. Attachment behavior of mammals. Psychol. Rev. 73:
409-426.

——, and D. L. JOHNSON, 1965. The development of interspecies social preferences. Psychon. Sci. 2:337-338.

FREEDMAN, D. G., J. A. KING, and O. ELLIOT, 1961. Critical period in the
social development of dogs. Science 133:1016-1017.

GOTTLIEB, G., 1965. Prenatal auditory sensitivity in chickens and ducks.
Science 147:1596-1598.

HERSHER, L., J. B. RICHMOND, and A. V. MOORE, 1963. Modifiability of
the critical period for the development of maternal behavior in sheep and
goats. Behaviour 20:311-320.

HUDSON, S. J., 1977. Multiple fostering of calves onto nurse cows at birth. Appl. Anim. Ethol. 3:57-63.

IMMELMANN, K., 1972. Sexual and other long-term aspects of imprinting in birds and other species. Adv. Study Behav. 4:147-174.

KILGOUR, R., 1972. Behavior of sheep at lambing. N.Z.J. Agric. 125:24-27.

KLOPFER, P. H., D. K. ADAMS, and M. S. KLOPFER, 1964. Maternal "imprinting" in goats. Proc. Natl. Acad. Sci. 52:911-914.

LORENZ, K. Z., 1937. The companion in the bird's world. Auk 54:245-273.

McBRIDE, G., I. P. PARER, and F. FOENANDER, 1969. The social organization and behaviour of the feral domestic fowl. Anim. Behav. Monogr. 2:127-181.

MORGAN, P. D., C. A. P. BOUNDY, G. W. ARNOLD, and D. R. LINDSAY, 1974. The role played by the senses in the ewe in the location and recognition of lambs. Appl. Anim. Ethol. 1:139-150.

NEATHERY, M. W., 1971. Acceptance of orphan lambs by tranquilized ewes (*Ovis aries*). Anim. Behav. 19:75-79.

SCHEIN, M. W., 1963. On the irreversibility of imprinting. Z. Tierpsychol. 20:462-467.

SCOTT, J. P., 1962. Critical periods in behavioral development. Science 138:949-958.

——, J. M. STEWART, and V. J. DeGHETT, 1974. Critical periods in the organization of systems. Dev. Psychobiol. 7:489-513.

SMITH, F. V., C. VAN-TOLLER, and T. BOYES, 1966. The "critical period" in the attachment of lambs and ewes. Anim. Behav. 14:120-125.

STANLEY, W. C., and O. ELLIOT, 1962. Differential human handling as reinforcing events and as treatments influencing later social behavior in Basenji puppies. Psychol. Rep. 10:775-788.

TYLER, S. J., 1972. The behavior and social organization of the New Forest ponies. Anim. Behav. Monogr. 5:85-196.

VINCE, M. A., 1973. Effects of external stimulation on the onset of lung ventilation and the time of hatching in the fowl, duck and goose. Br. Poult. Sci. 14:389-401.

WOOLPY, J. H., and B. E. GINSBURG, 1967. Wolf socialization: a study of temperament in a wild social species. Am. Zool. 7:357-363.

WYATT, R. D., M. B. GOULD, and R. TOTUSEK, 1977. Effects of single vs. simulated twin rearing on cow and calf performance. J. Anim. Sci. 45:1409-1414.

10

Learning

Most ethologists agree with Thorpe (1963) that learning has occurred when an individual behaves in a more adaptive way because of experience. Three salient points about the process, when viewed in this way, are:

1. A *change in overt behavior* indicates that learning has occurred.
2. *Experience is responsible* for the change.
3. *The change benefits the individual.*

Some would argue that learned behavior is not necessarily adaptive. However, nonadaptive behavioral changes are more likely in artificial than in natural environments. One aspect of a study by Schake and Riggs (1972) may be used to illustrate this. They brought in beef cows from several different sources (all had been on range) and placed them in an open-lot, cow–calf confinement system of management. Cows were confronted with strangers in a crowded environment and with the necessity of competing for feed. During the first 4 days, high levels of fighting, butting, and threatening were observed. As a result, more than 10% of the cows became so submissive that they would not compete for feed. Subordinate cows became submissive; they learned to remain away from the feeder while their more dominant herdmates fed. Was their learned behavior adaptive? The immediate result of remaining away from the feeder was advantageous, as they avoided further punishment and possible injury; the longer-term result was that they were becoming weaker and might have starved had the experimenters not fed them separately for several weeks until the frequency and severity of agonistic activity decreased. Under more natural range conditions, a cow threatened by a dominant herdmate would simply withdraw and search for food elsewhere. In the confinement system, food was not available elsewhere.

SPECIES AND STRAIN DIFFERENCES

It is clear that capabilities for learning vary greatly among species when the entire animal kingdom is considered. From the evolutionary viewpoint, animals can be classified as to their general level of intelligence (ability to learn). Thus, worms are inferior to fish, and amphibians are inferior to mammals. However, animal learning ability shows a high degree of specialization; ecological adaptation plays a major role in determining what can be learned and how easily (Bolles, 1970). For example, honeybees, which are low on the taxonomic scale of domestic animals, show much instinctive behavior. Nevertheless, if a hive is moved to a new location, foraging workers learn rapidly where fields of flowers are and the location of their hive. Precocial birds, lower on the scale than mammals, learn to follow their mother shortly after hatching. On the other hand, altricial mammals, such as the carnivores and primates, which are high on the evolutionary ladder, receive much maternal care during early life and learn little during the first weeks or months. However, they eventually learn a great deal and the behavior that is learned varies greatly among individuals.

Which is the most intelligent farm animal? Consider what the ability to learn a maze indicates about relative intelligence. Psychologists have conducted many studies involving rats in mazes. This species is well suited for learning how to get through. Rats have great exploratory drive and without any obvious reward will familiarize themselves with tunnels and complex passageways. Learning how to get through maze-like passages is adaptive to a rat, as it often provides access to food and shelter from predators. What would it mean if horses, cattle, pigs, sheep, dogs, and chickens were compared for intelligence in mazes of similar complexity but scaled according to the size of the animal? Sheep and chickens, frightened by being isolated from their companions and by the strange surroundings, might "freeze" for a long period; beef cattle and pigs might respond by lying down and sleeping; horses and dogs might run about wildly or trot through undisturbed, depending on their breed or on their previous experience. Speculation would only be replaced by facts by eventually doing the experiment. Those responses could be fascinating and suggest other studies. Much variability would probably be found among individuals, families, and breeds within species. But what would have been learned about relative intelligence? Motivation levels, emotionality, sensory capabilities, agility, and other characteristics of the various species could be confounded with what is called learning ability in terms of getting through a maze.

Genetic differences produced by man's selection may have pronounced effects on the ability of different stocks within a species to learn specific things. Evidence was cited in Chapter 7 showing that artificial selection was effective in separating strains of rats for ability to learn mazes. However, the responses caused by selection in one such pair of lines was found to be for the particular kind of maze in which selection was conducted; "maze-bright" rats learned better in situations where they were rewarded by food, and "maze-dull" rats did better in learning to escape from water (Searle, 1949). Similarly, some breeds of dogs are better than others in solving problems requiring independent action, whereas a different set of breeds excel in obedience training (Scott and Fuller, 1965).

HABITUATION

When an animal stops responding to a specific, neutral stimulus, it has become habituated or extinction has occurred. The distinction will be maintained here that habituation refers to the loss of instinctive responses and extinction to the cessation of learned behavior. Habituation is perhaps the simplest kind of learning; it is learning *not* to respond. Nevertheless, it is of great importance to the animal, as it reduces the use of energy for nonadaptive activity.

The critical period for socialization of the young animal typically comes to an end as fear of novel objects and situations increases (Figure 9-1). The devlopment of fear to that which is strange is of general value to a young animal after it has formed its initial social bonds; such fear keeps the young close to the dam for "solitary" birds and mammals and to flock or herdmates for social animals. Staying with the dam or group provides greater protection from predators, food availability, the shelter of other bodies in cold environments, and the opportunity to learn many behaviors.

As the young animal develops further, it begins to lose many of its fear responses by habituation. Withdrawal and escape reactions to specific stimuli that attract the animal's attention but are otherwise harmless are not worth the effort; there is no reward for fearful responses and no punishment for lack of such a response. Habituation to harmless, but initially fear-producing stimuli continues throughout an animal's life, but we expect this to be less important to older animals in a relatively unchanging environment.

Habituation or loss of instinctive response also occurs to other than fear-producing stimuli (e.g., to nonfood items in the young animal). Thus, newly hatched chicks and poults peck at small objects that contrast with the background, but they soon stop pecking nonfood objects as they learn to identify edible food. Similarly, mammalian young usually stop nonnutritive sucking soon after locating milk-producing nipples.

It should be noted that habituation is not necessarily permanent and has some characteristics in common with other kinds of learning (Thompson and Spencer, 1966). For example, if habituation has been established to a particular stimulus and that stimulus is withdrawn for a sufficient interval, it may again cause the animal to respond when it is presented at a later time. Also, after an animal is habituated to a particular stimulus, it is likely to show little or no response to similar stimuli.

ASSOCIATIVE LEARNING

Stimuli involved in the learning process generally fall into one of two categories: they are associated either with rewards (positive reinforcers) or with aversive effects or punishment (negative reinforcers). As discussed in the previous section, objects or situations that have neutral effects may initially act as stimuli, but as the animal habituates to them, they lose their stimulating properties.

Learning brought about by positive or negative reinforcers is known as *associative learning*. Extensive laboratory studies of learning processes have produced a variety of classifications (see Dewsbury, 1978:321-331, for a summary). Two major kinds of associative learning are now recognized in laboratory settings: they are usually called *instrumental* or *operant conditioning* and *classical conditioning*. Learning under less artificial conditions often involves trial and error and is similar, in many respects, to instrumental or operant conditioning as studied in the laboratory. The essential difference between operant and classical conditioning is that the animal has some control over what happens under operant conditioning, whereas under classical conditioning it does not.

PROBLEM SOLVING
AND OPERANT CONDITIONING

Domestic animals are able to solve many problems of importance to them in a natural setting. Among problems encountered are these: What is appropriate food, and where is it located? Where is water located? What companionship is available and desirable? How can temperature be regulated for comfort? What objects and animals are to be avoided?

Behavior can sometimes be used as a guide in making management decisions (e.g., to indicate those foods and environmental conditions preferred by the animal). Little is known as to how domestic animals develop early food preferences, but studies with rats suggest that two kinds of experience influence those preferences (Galef, 1976). Thus, nursling rats are influenced by the diet of the mother (from taste cues in the milk), so that they later recognize and prefer food items eaten by their dam during lactation. Also, young rats usually feed in the presence of adults and thereby learn to eat the same foods. The same mechanisms for learning food preferences may be important in domestic animals that have a choice of foods available, as for animals on pasture where several species of plants are present. Temperature and light preferences of pigs have been investigated by training pigs to control those variables, for example by breaking a photoelectric beam to obtain short periods of exposure to heat or to light (Baldwin and Meese, 1977).

Operant Conditioning: An Example

Investigators at the Ruakura Animal Research Station in New Zealand (Moore et al., 1975) suggested that dairy cows might be allowed to manipulate their own environment (such as vacuum and pulsation rates of milking machines), feed, temperature, humidity, music, and so on, after being trained by operant conditioning techniques. Such a procedure would allow researchers to determine preferences, and perhaps to provide in this way an environment resulting in higher productivity. In addition, Moore et al. hoped to gain insight into the general ability of cattle to learn. Therefore, an experimental chamber was set up, containing a "nuzzle plate" which could be pushed and a feed delivery system which was activated electronically when the plate was pushed by the cow's muzzle (Figure 10-1). Delivery of feed could be controlled according to desired ratios, such as delivery of feed after being pushed 10 times. Table 10-1 presents some of the data from that study, which help to illustrate the methods employed and results obtained.

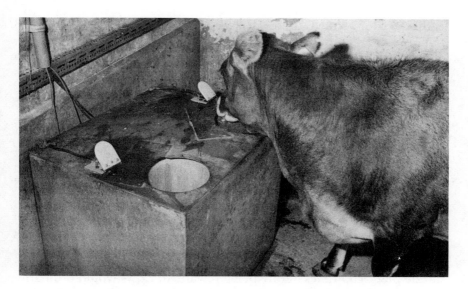

Figure 10-1 A hungry cow pushes a nuzzle plate, causing feed to be delivered. Animals can be trained by instrumental or operant conditioning methods to control their environment so as to satisfy needs or to indicate preferences. (Courtesy of Ron Kilgour, Ruakura Animal Research Station, New Zealand.)

Table 10-1 *Number of sessions required to habituate cows to the experimental chamber, to shape the operant behavior, and to reach a fixed ratio of 100:1*

	Sessions required		
Cow number	Habituation	Training (shaping)	To attain ratio of 100:1 (days)
683	0	1	18
041	3	4	16
566	4	8	27
663	6	5	21
7153	4	4	24
7102	4	14	25
2	10	12	14

SOURCE: Modified from Moore et al. (1975).

Cows were motivated to respond to the positive reinforcer (feed) by being kept hungry for an extended period prior to being placed in the chamber. They were left in the chamber for 25-minute periods. Habituation was indicated by cows readily entering the chamber and eating feed provided there (fear responses disappeared). Cow 683 required no habituation period because the experimental chamber was a part of her familiar environment and she was accustomed to eating the concentrated feed provided. All other cows were strangers to the environment and were not accustomed to being fed concentrates.

Training or "shaping" of the operant response followed the method popularized by Skinner of rewarding closer and closer approximations of the desired behavior. Thus, during the early part of the training period, cows had feed delivered for approaching and moving their heads near the nuzzle plate. Later, feed was delivered each time the nuzzle plate was touched.

Different schedules of reinforcement were tested after individual cows learned to operate the feeding device. After the initial training period, feed was delivered on fixed or variable schedules. For example, in the case of the fixed ratio 100:1, a cow would have to push the nuzzle plate 100 times before being rewarded. Cows would ordinarily push the plate rapidly until fed. Hungry cows will work hard to obtain feed. The Ruakura workers concluded that cows can readily learn to manipulate their environment using techniques found to be effective in other species.

CLASSICAL CONDITIONING

The Russian physiologist Pavlov is regarded as the "father" of classical conditioning. Pavlov stumbled onto this learning process while studying the physiology of digestion. His dogs, which had cannulas (small tubes) surgically implanted in their salivary glands and exteriorized, were observed to begin secreting saliva as soon as he entered the room and before they were fed. Pavlov spent much of his life studying related phenomena. A large number of psychologists followed his lead with extensive studies of the classical or Pavlovian conditioning process. They believed that the results were likely to be applicable not only to their experimental animals, but to a wide variety of higher animals, including man.

A typical Pavlovian experiment was as follows (Pavlov, 1927). A hungry dog with a tube inserted into its salivary gland was found to salivate when meat powder was placed in its mouth. The meat powder was termed an *unconditioned stimulus* (US) and the salivary response an *unconditioned response* (UR). An ethologist might use the terms "releaser" and "instinctive behavior" for the meat powder and salivary response, respectively. In contrast to the meat powder, the

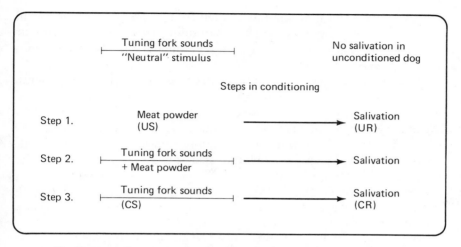

Figure 10-2 Classical conditioning of the salivary response in a dog. Conditioning occurs as a previously neutral stimulus becomes effective in causing a response in a motivated animal. Pairings of unconditioned stimulus (a natural releaser) and of the intended conditioned stimulus are required for the animal to learn to respond in this way.

tone produced by a tuning fork was neutral with respect to salivation. Pavlov "conditioned" the animal to salivate in response to the previously neutral stimulus (the tone) by first sounding the tone and then putting some meat powder into its mouth. After about 10 pairings, the dog would produce a small amount of saliva in response to the tone alone; after 30 pairings, it would salivate profusely after the tone was heard. Following the learning process the tone was termed a *conditioned stimulus* (CS), and salivation in response to the tone alone was called a *conditioned response* (CR). We may visualize this experiment as shown in Figure 10-2.

The choice of stimuli to be used for conditioning animals is important. If habituation to a stimulus has already occurred, it is difficult at a later time to use the same signal as a conditioned stimulus; a novel and attention-getting stimulus works better.

Generalization, Discrimination, and Extinction

A dog conditioned to a particular tone will generally respond in the same way to a similar, but different tone. This generalization of response to a class of stimuli is an important characteristic of learning.

Horses, dogs, and other animals trained for performance or work will generally show generalization by responding to the same or similar commands or signals given by more than one person, although conditioned by only one person.

Discrimination between similar stimuli is the opposite of generalization. Another experiment by Pavlov illustrates the difference between generalization and discrimination. He conditioned a dog to salivate by pairing meat powder with a metronome ticking at 100 beats per minute. After the conditioned response was well developed, Pavlov found that the dog would also salivate in response to the metronome beating at the rate of 80 per minute (generalization), but not when the rate was 60 per minute (discrimination). Generalization and discrimination testing have proven useful in testing sensory perceptions in animals.

After an animal has been conditioned, repeated presentation of the conditioned stimulus without reinforcement generally leads to loss of the conditioned response (i.e., extinction occurs). Extinction is not necessarily permanent, however, as the animal will usually show the conditioned response again (although it may be weakened) after a "rest" period; this phenomenon was called *spontaneous recovery* by Pavlov.

Timing Relationships

Psychologists and animal trainers are well aware of the importance of rather precise timing in pairings of the unconditioned and intended conditioned stimulus. Numerous studies indicate that the intended conditioned stimulus should occur before the unconditioned stimulus. Usually, the interval between conditioned and unconditioned stimuli must be very brief for effective conditioning. Generally speaking, the delay must not be more than a few seconds. A similar relationship holds in operant conditioning. For example, Moore et al. (1975) devised their apparatus for conditioning cows so that feed was delivered (during training) less than 2 seconds after the nuzzle plate was pushed.

Delayed Reinforcement and Taste Aversions

In most associative learning it has been established that the reward or punishment (the unconditioned stimulus) must be closely associated in time with the previously neutral stimulus if that stimulus is to produce a conditioned response, as described above. However, some important exceptions have been found. For example, it has been learned that rats develop taste aversions to substances that have a different taste and are followed by illness beginning as much as an hour later (Garcia et al., 1966). This may explain why rats readily become "bait-shy"; they are prepared by their physiology and life-style to eat a large variety of foods, but food that tastes different is eaten in limited quantity and avoided later if illness follows ingestion. In contrast, quail develop aversions to foods if they look different and later cause illness (Garcia et al., 1974), as quail are more visually oriented than rats. An interesting possibility is that coyotes can be conditioned to avoid killing and eating sheep by putting out baits containing the flavor of lamb or mutton but laced with lithium chloride, which makes the predator ill after a time delay (Gustavson et al., 1976).

Strength of Stimulus and Rate of Learning

Stimuli must be of sufficient strength to attract the animal's attention, and unconditioned stimuli must cause the appropriate response if animals are to be conditioned in ways desired by man. Both positive and negative reinforcers (unconditioned stimuli) are used in training animals. Although positive reinforcers are generally more effective when strong, there are indications that negative or aversive stimuli may cause emotional upset and fear when overly intense, so that inappropriate responses occur and learning of desired responses may be slow or absent. As an example, in laboratory settings it has been found that electric shocks applied to rats for avoidance conditioning were more effective with increasing intensity up to moderate levels and that higher intensities caused crouching or "freezing" rather than avoidance (Moyer and Korn, 1964). It appears that excessive strength of aversive stimuli in conditioning animals (e.g., extreme application of spurs, prods, bit pressure, and choke chains) may be counterproductive, as excessive emotionality and fear interfere with the desired responses. Although negative reinforcement is commonly and effectively used in training animals, it can obviously be carried too far. The trainer himself should avoid becoming a conditioned stimulus eliciting fearful responses, unless that is his specific objective.

Conditioning Schedules

There appears to be considerable variation among species, but dog and horse trainers generally believe that the number of pairings of conditioned and unconditioned stimuli per session should be relatively few and that training sessions should occur a day or more apart. Identification of superior reinforcement schedules for domestic animals for different kinds of learning is a desirable goal for future study.

Karas et al. (1962) demonstrated the effect of spacing learning sessions in 3-week-old pigs. They trained pigs to avoid receiving an electric shock by jumping across a small board into an adjacent compartment when warned by the sound of a buzzer continuing for 6 seconds before the shock was delivered. A pig could avoid being shocked by responding within the 6-second interval but was otherwise punished by the shock. After a brief interval, the buzzer would sound again, and the pig was then required to jump back into the original compartment to avoid another shock on the other side of the apparatus. The experimenters chose three conditioning schedules, consisting of 40 pairings each. One group of pigs was given the 40 trials during a brief interval all on the same day (the pig was required to jump back and forth 40 times within the same learning session). A second group was given 20 trials per day on two consecutive days. The third group had 10 trials daily on four successive days. Karas et al. then compared the percentage of avoidances during each of four sets of 10 consecutive trials for each of the three groups of animals, with results as shown in Figure 10-3. During the final 10 of 40 trials, the three groups of pigs performed at very different levels; those given massed pairings (40 trials during a single session) were successfully avoiding the shock less than 20% of the time, those given two sessions avoided shocks less than 50% of the time, and those given four sessions of 10 pairings each avoided being shocked on more than three-fourths of all trials during the final session.

Further studies were carried out with Duroc and Hampshire pigs by Willham et al. (1964) using essentially the same procedures. They found that nine daily sessions of 10 pairings each resulted in well over 90% avoidances without shock during the final session. The learning curve was nearly linear during the first four sessions, but further increases in improvement became more and more difficult to achieve. It was also established that the Duroc breed learned at a significantly faster rate than the Hampshire breed during the first few sessions, but both breeds eventually achieved the same level of success. Although the reason for the breed difference was not established experimentally, it was suggested that the Duroc breed is more docile and less emotional than the Hampshire and therefore more easily taught by a negative reinforcer (electric shock), as they become less emotionally upset during training sessions.

How can the effectiveness of conditioning or training be eval-

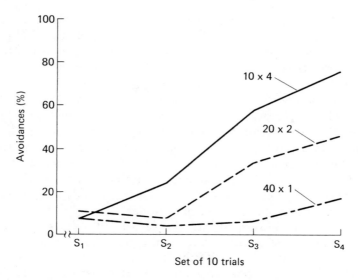

Figure 10-3 The relative effectiveness of spacing training sessions as indicated by percentages of successful avoidance of electric shocks following a 6-second warning by a buzzer. Three-week-old pigs were far more successful in learning to avoid when given four sessions of 10 pairings each than when given 40 pairings in one long session; pigs given two sessions of 20 pairings each showed an intermediate response. (Reprinted with permission of authors and publishers from: Karas, G. G., Willham, R. L., and Cox, D. F. Avoidance learning in swine. Psychological Reports, 1962, 11, Figure 1, Page 52.)

uated? Rate of learning is one measure of success and extinction rate is another. Usually it is preferable that an animal should have a slow rate of extinction. However, extinction may occasionally be useful to an animal by allowing it to stop responding in a way that no longer satisfies a need or drive because of a changed situation. Some techniques that lead to more rapid learning (acquisition) are also associated with rapid extinction, and slow acquisition because of irregular or weak reinforcement may be offset by slower extinction (Tarpy, 1975:140-154).

LEARNING AND NEUROTIC BEHAVIOR

Increasingly difficult discrimination testing can lead to neurotic behavior (Pavlov, 1927). Dogs were given a problem in which they had to make discriminations between circles of light (resulting in a food reward) and ellipses (no reward). As the experiment proceeded, the ellipse was changed to more closely approach the shape of a circle. As discrimination became progressively more difficult, the dogs showed obvious signs of emotional upset; they became unruly and made desperate attempts to escape.

Immobility may be associated with a neurotic response (Kurtsin, 1968:80-82). A hungry cat, accustomed to killing and eating mice, was wired for electric shock. The cat received a punishing shock when it pounced on a mouse introduced into its cage. After attempting to escape, the cat became immobile. In the following weeks the cat refused to approach mice. Although remaining immobile while mice crawled over it, the cat was obviously fearful, as indicated by abnormal electrocardiograms during such episodes.

Simultaneous exposure of cats to positive and negative reinforcers was found by Masserman (1943) to produce neurotic behavior. Hungry cats were trained to lift the lid of a food box and feed when a sound or light signal was presented. After being well conditioned in this way, they received a blast of air in the face while feeding (cats dislike this treatment). Neurotic behavior was produced, including restlessness, trembling, feeding inhibition, and attempts to escape. Neurotic behaviors were exaggerated when the light or sound was presented. Although normal cats normally do not drink milk containing alcohol, Masserman's neurotic cats preferred the mixed drink and showed fewer symptoms when intoxicated.

EARLY EXPERIENCE

Studies with laboratory animals (especially rats) indicate that stressful conditions encountered by the mother during pregnancy can have significant effects on the "emotionality" of the offspring produced, even after the progeny mature (Thompson, 1957). It appears that the effect is mediated by hormones produced by the dam during stress (e.g., adrenocorticotropic hormone and cortisone, which pass into the bloodstream of her fetuses).

The effects of early experience on emotionality can be complex, and emotionality can alter ability to learn, particularly when negative reinforcers are involved. Denenberg (1969) has postulated that there is an optimal range of stimulation in infancy. Evidence cited by Denenberg indicates that what is optimal for one species or genetic stock may not be optimal for another. This implies that persons interested in the learning ability of domestic animals may need to determine the effects of particular levels of infantile stimulation on emotionality of their breed or strain and proceed accordingly.

A study by Denenberg and Karas (1960) illustrates how emotionality influenced by infantile experience affects learning ability at a later age. Their stimulation consisted of simply handling individual rats for brief daily periods and then testing at a later age for speed of learning to avoid shock following a signal. Control animals, not handled prior to weaning at 21 days, were excessively emotional and overresponded to the signal; in their panic they had great difficulty in finding the escape box. Rats handled for the first 20 days after birth were also poor learners, but for a very different reason; they were relatively nonemotional, and some would sit on the electrified grid temporarily before moving about, slowly seeking an exit. Animals handled for either the first 10 days after birth or the last 10 days before weaning (at 21 days) were intermediate in emotionality and easily learned to avoid being shocked in response to the signal.

The ability of dogs to learn later in life as the result of "restricted" compared to "enriched" rearing environments was studied in Scottish Terriers by Thompson and Heron (1954). They subdivided puppies from the same litters at weaning so that half lived as individuals, were exposed to infrequent human contact, and had minimal to moderate sensory input (restricted). The other half were placed in homes where human contact was friendly and frequent and the animals had daily exercise (enriched). At 8 months of age, all animals were returned to the laboratory, where they were then tested for ability to solve a variety of problems. Differences between the two groups were present and persisted for the entire test period (beyond 18 months of age). Problem solving by dogs reared in deprived environments was always strikingly inferior to that of dogs exposed to enriched environments during rearing. They showed that older dogs could easily learn new tricks if reared in a stimulating environment from 2 until 8 months of age; otherwise, they were grossly deficient.

TRAINING ANIMALS

That higher animals can be trained to carry out a variety of complex behaviors is evident from the performances of circus animals, riding horses, hunting dogs, guide dogs for the blind, and the like. Nevertheless, the amount of time involved may be great, and animals (and their trainers) vary greatly in their abilities. Some basic principles of learning have been indicated in this chapter, and the concepts of motivation and socialization are also involved in important ways in an animal's learning when controlled by man (see Chapters 8 and 9). Training animals to complex tasks requires considerable skill and patience. The assistance of an experienced animal trainer or close study of specialized books on training should be very helpful to those who would like to become experts themselves. Some useful books for horse and dog training have been written by Miller (1975), Pfaffenberger et al. (1976), and Editors of Sports Illustrated (1972).

A few examples of animal training will now be examined to reemphasize principles already covered.

Training Dairy Heifers to Lead by Two Methods

It is desirable that dairy heifers be taught to lead in response to minimal tension on a hand-held halter rope. Wilson et al. (1975) have compared the conventional method of tying and leading by hand with a modification of the so-called Australian "bullock"

Figure 10-4   Heifers learning to be led by the "bullock method." The animal wearing the collar has the mechanical advantage and can force the other to move about or to remain close by. By reversing the leader–follower pairing daily, for a few days, each animal learns to be led when wearing the halter. Most heifers are easily led after such training; they generalize and follow either another heifer or a person when led by a halter rope. (Courtesy of James Wilson, Hawkesbury Agricultural College, Richmond, N.S.W., Australia.)

method. Dairy heifers are commonly socialized to man. Nevertheless, they may be frightened of halters and resist being led. After becoming habituated to the halter and lead rope by being tied, they can be led about with the use of some force if small enough to manage (usually as calves less than 6 months old). Older animals may require more force than can be exerted by a man on foot, so that they must be pulled about initially by a tractor or truck; most farmers would not consider this to be a practical alternative.

The Australian bullock method involves tying heifers together by pairs matched for size as shown in Figure 10-4. One heifer wears a leather collar and the other a halter; the collar and halter are attached by a short chain (about 20 inches long). The mechanical advantage belongs to the animal wearing the collar, so that it determines where the pair goes. During the first 2 days the leader–follower pairing is reversed twice a day; thereafter reversals are made daily. After 7 or 8 days of such training, most heifers can easily be led by a person. Those not responding were continued in pairings up to 21 days. A few recalcitrant animals may never learn to be led by man. Nevertheless, the method is generally satisfactory (although some animals get a leg over the chain during the first day or two; this is easily corrected). When a number of pairs were trained to lead during the same period, only about 1 hour of cumulative time was required for training each heifer by this method, as compared to $3\frac{1}{2}$ hours by the conventional procedure. The heifers showed generalization; they were as easily led by a person as by another heifer after learning to respond to a minimum pull on the lead rope.

Training Horses for Pulling Contests

Horses used for pulling contests commonly receive training believed to maximize rate of acquisition and to delay extinction (Kratzer, personal communication). Although methods may differ somewhat, the following is used by some trainers. Initially, training involves sounding of a buzzer (conditioned stimulus) followed by an electric prod (unconditioned stimulus), which causes an escape reaction (unconditioned response). The escape reaction is a powerful one, and the horse "hits" the load with great force. The buzzer and prod are paired many times (numerous training sessions) with continuous reinforcement. During the second phase of training an irregular schedule of aversive reinforcement is used (i.e., the prod is used erratically after the buzzer, so that the horse never knows when it may occur). The final phase takes place during contests. At that time the buzzer sounds and the horse hits the load. Because prods are not allowed during contests, the response given then is truly a conditioned response. The trainer relies on a slow extinction rate during contests.

Some Factors in Trainability of Guide Dogs

Pfaffenberger et al. (1976) have outlined the state of the art as applied to the selection, development, and training of guide dogs for the blind. Their program, as applied to the individual prospective guide dog, includes a period of individual handling (6 to 8 weeks), puppy testing (8 to 12 weeks), and home training (12 weeks to 10 to 15 months) which precedes the final intensive training period. Pfaffenberger and Scott (1959) discovered that a major shortcoming during early years of the program was inadequate socialization of the puppy to man and restricted early experience. The main results of that study are presented in Table 10-2. It is clear that continued social contacts are essential during the juvenile period of development.

An important practical consideration in training animals is the individual's "trainability," which is usually associated more with emotionality than with differences in intelligence per se, at least in dogs. Pfaffenberger and his colleagues devised an early testing procedure, applied from 8 to 12 weeks of age, which serves a double purpose; the puppies are not only screened for future trainability but are also exposed to people and varied experience which increases the socialization process and "enriches" early experience. In a limited sample it was found that about 60% of pups passing the screening test succeeded in later guide-dog training as compared to 20% percent success of those that failed in the early testing program.

Table 10-2 *Trainability of dogs for guiding the blind as affected by retaining puppies in kennels beyond 12 weeks[a]*

Placed in 4-H homes within:[b]	Number of puppies succeeding/ total in guide-dog training	Percentage becoming guide dogs
1 week	36/40	90
2 weeks	19/22	86
3 weeks	11/19	58
>3 weeks	13/43	30

[a]Puppies had about 30 minutes of close contact with people weekly from 8 to 12 weeks of age.

[b]Interval between last week of close human contact (30 min/week) and arrival in 4-H home.

REFERENCES

BALDWIN, B. A., and G. B. MEESE, 1977. Sensory reinforcement and illumination preference in the domesticated pig. Anim. Behav. 25:497-507.

BOLLES, R., 1970. Species-specific defense reactions and avoidance learning. Psychol. Rev. 77:32-48.

DENENBERG, V. H., 1969. The effects of early experience. In E. S. E. Hafez, The Behaviour of Domestic Animals, 2nd ed. The Williams & Wilkins Company, Baltimore.

——, and G. G. KARAS, 1960. Interactive effects of age and duration of infantile experience on adult learning. Psychol. Rep. 7:313-322.

DEWSBURY, D. A., 1978. Comparative Animal Behavior. McGraw-Hill Book Company, New York.

EDITORS OF SPORTS ILLUSTRATED, 1972. Sports Illustrated Dog Training. J. B. Lippincott Company, Philadelphia.

GALEF, B. G., 1976. Social transmission of acquired behavior: a discussion of tradition and social learning in vertebrates. Adv. Study Behav. 6:77-100.

GARCIA, J., F. R. ERWIN, and R. A. KOELLING, 1966. Learning with prolonged delay of reinforcement. Psychon. Sci. 5:121-122.

GARCIA, J., W. G. HANKINS, and K. W. RUSINIAK, 1974. Behavioral regulation of the milieu interne in man and rat. Science 185:824-831.

GUSTAVSON, C. R., D. J. KELLEY, M. SWEENEY, and J. GARCIA, 1976. Prey–lithium aversions; I. Coyotes and Wolves. Behav. Biol. 17:61-72.

KARAS, G. G., R. L. WILLHAM, and D. F. COX, 1962. Avoidance learning in swine, Psychol. Rep. 11:51-54.

KURTSIN, I. T., 1968. Pavlov's concept of experimental neurosis and abnormal behavior in animals. In M. W. Fox (Ed.), Abnormal Behavior in Animals. W. B. Saunders Company, Philadelphia.

MASSERMAN, J. H., 1943. Behavior and Neurosis. University of Chicago Press, Chicago.

MILLER, R. W., 1975. Western Horse Behavior and Training. Dolphin Books, Doubleday & Company, Inc., Garden City, N.Y.

MOORE, C. L., W. G. WHITTLESTONE, M. MULLORD, P. N. PRIEST, R. KILGOUR, and J. L. ALBRIGHT, 1975. Behavior responses of dairy cows trained to activate a feeding device. J. Dairy Sci. 58:1531-1535.

MOYER, K. E., and J. H. KORN, 1964. Effect of UCS intensity on the acquisition and extinction of an avoidance response. J. Exp. Psychol. 67:352-359.

PAVLOV, I. P., 1927. Conditioned Reflexes. Translated by G. V. Anrep. Oxford University Press, London.

PFAFFENBERGER, C. J., and J. P. SCOTT, 1959. The relationship between delayed socialization and trainability in guide dogs. J. Gen. Psychol. 95: 145-155.

——, J. L. FULLER, B. E. GINSBURG, and S. W. BIELFELT, 1976. Guide Dogs for the Blind: Their Selection, Development, and Training. Elsevier Scientific Publishing Co., New York.

SCHAKE, L. M., and J. K. RIGGS, 1972. Behavior of beef cattle in confinement. Tex. Agr. Exp. Sta. Tech. Rep. 27.

SCOTT, J. P., and J. L. FULLER, 1965. Genetics and the Social Behavior of the Dog. University of Chicago Press, Chicago.

SEARLE, L. B., 1949. The organization of hereditary maze-brightness and maze-dullness. Genet. Psychol. Monogr. 39:279-325.

TARPY, R. M., 1975. Basic Principles of Learning. Scott, Foresman and Company, Glenview, Ill.

THOMPSON, R. F., and W. A. SPENCER, 1966. Habituation: a model phenomenon for the study of neuronal substrates of behavior. Psychol. Rev. 73: 16-43.

THOMPSON, W. R., 1957. Influence of prenatal maternal anxiety on emotionality in young rats. Science 125:698-699.

——, and W. HERON, 1954. The effects of restricting early experience on the problem-solving capacity of dogs. Can. J. Psychol. 8:17-31.

THORPE, W. H., 1963. Learning and Instinct in Animals, 2nd ed. Methuen & Company Ltd., London.

WILLHAM, R. L., G. G. KARAS, and D. C. HENDERSON, 1964. Partial acquisition and extinction of an avoidance response in two breeds of swine. J. Comp. Physiol. Psychol. 57:117-122.

WILSON, J. C., J. L. ALBRIGHT, J. L. COLLINS, G. BUGDEN, A. EDEN, and R. J. BUESNEL, 1975. Training heifers to lead in pairs. J. Dairy Sci. 58:749 (Abstract).

11

Aggression

Aggressive behavior involves some form of threat or attack, usually directed toward another member of the same species. Attacks of predators on their prey are excluded from consideration here; such acts by predators are more logically included as part of the appetitive phase of ingestive behavior.

The primary role of aggression in natural environments is to assure an adequate supply of scarce resources to high-status animals when competing with their own kind. It appears reasonable to assume that nature must exert some selection pressure for aggressive behavior. However, further consideration suggests that natural selection is for an optimal level of aggressiveness rather than an extreme. As the young animal develops, it would, in fact, be disadvantageous for it to be overly aggressive. Older and stronger animals are likely to rebuff an aggressive youngster that attacks and may injure it if it persists. Under those circumstances the ability to learn submissive or appeasement behavior in the presence of a more powerful and aggressive animal is also a component of fitness.

Although social organization of a group may, in the long run, be advantageous for its members, and although some aggressive acts may be necessary in establishing and maintaining the order, excessive aggression can be harmful in terms of wasted energy and possibly serious injuries. Influences that delay the onset and severity of aggressive behavior are worthy of study, as are those that predispose animals to aggression or increase the frequency of social conflict.

DELAYED AND REDUCED AGGRESSION

Long-lasting social tolerance and preferences for particular individuals usually result from exposure of young animals to those individuals during very early life. However, social attachments can range from the highly specific mother–young pairing to the very general case of species recognition and preference. There is evidence that young animals often form strong attachments for each other. This is especially likely when man severs the natural mother–young bond at an early age or prevents it from forming altogether. Schloeth (1961) observed that young calves sleep together during the greater part of the day, while their mothers graze and only come to nurse their calves at certain times. When traveling or fleeing, calves were seen to move in close-up groups. However, he observed in horses that foals stayed close to their individual dams when groups showed escape reactions.

Twin calves usually remain close together while grazing on pasture. Ewbank (1967) investigated this spacing relationship further by also keeping unrelated calves together by pairs in pens during rearing (as twins were kept). Unrelated calves reared together as pairs from 2 weeks old showed the same preference for each other's close company while grazing on pasture, as did twin calves that had been kept together from birth in the same manner. It appears that affiliative behavior of young cattle is controlled more by social rearing conditions than by genetic likeness.

Bouissou and Andrieu (1978) found that calves kept together from birth form stronger preferences for each other and show greater social tolerance than do calves brought together for the first time at 6 months of age or older and then tested at later ages. Those brought together at birth were combined in similar subgroups at 6 months. After 12 months in the larger group, those grouped at birth showed only about half as many agonistic acts and more than five times as many affiliative, nonagonistic acts, such as smelling, licking, elements of sexual behavior, and play toward their constant companions as toward those with whom they had been assembled at 6 months of age. Animals raised together from birth exhibited much less aggressive behavior toward each other when feeding together.

As group-living animals mature, their social tolerance generally decreases, agonistic activity begins, and social hierarchies form. Even so, groups kept together from hatching or birth begin to show aggressiveness and become socially organized at later ages. Guhl (1958) found that isolated cockerels assembled at 31 days of age established most of their dominance relationships within 6 days, which was much earlier than for group-reared cockerels. In a somewhat similar study, Bouissou and Andrieu (1977) found that strange heifer calves brought together at 6 months formed dominance-subordinance relationships promptly, but heifers kept together from birth did not develop such relationships until about 9 months of age. Forcing competition between calves for feeding privilege at an artificial teat caused a feeding-right dominance for calves that were only a few weeks old (Stephens, 1974), but the consequences beyond 9 weeks were not studied. On the other hand, Schake and Riggs (1970) did not find social hierarchies in groups of calves raised together from birth until 325 days. From these studies it is clear that early and continuous association is associated with greater social tolerance, delayed onset of aggressive behavior, and relatively slow formation of social hierarchies.

MALE SEX HORMONES

Onset of Aggressiveness

As indicated earlier, adult males of most domesticated species are larger and more powerful than females and are better adapted for fighting or bluffing. Secondary sexual dimorphism is associated with the presence of functioning testes. This has been recognized since ancient times; castration has long been used to increase the docility of males of large and powerful animals such as horses and cattle. More recently, male sex hormones or androgens have been recognized as the physiological agents produced by the testes that are responsible for masculinization and the increased tendency for aggressive behavior of males.

The appearance and behavior of young chicks helps to demonstrate the role of male sex hormones. By about 4 or 5 weeks of age male chicks can be identified by the enlargement and reddening of their combs. During this same period there is considerable sparring (pretend fighting, without physical contact) and then aggressive pecking begins. Most of the early pecking is done by males. Guhl (1958) found that by 6 weeks all chicks in two heterosexual flocks could be sexed by their appearance (primarily comb size and redness), and up to this age those chicks identified as males had pecked at other chicks more than nine times as frequently as females.

In a further series of comparisons, Guhl (1958) subdivided 40 cockerels equally into four groups and then had three of those groups castrated at 9 days. Injections of androgen into two of the groups and frequent observation of all groups revealed the mean age at which *peck-rights* were established (see Table 11-1). A peck-right is present when one individual of a pair consistently avoids or behaves submissively when threatened or attacked by the other, thereby reducing the tendency of the more aggressive chick to continue its aggressive behavior. Although untreated capons were observed to deliver some pecks relatively early, they were of such low intensity and frequency that the mean age of peck-right formation occurred about 4 weeks later than in the intact male group. In contrast, the two capon groups deprived of their own testicular androgen but injected with large amounts, developed peck-rights even earlier than the intact males.

The role of male sex hormones in causing aggressiveness (or lowering the threshold for such behavior to occur when a stimulus object appears) is fairly evident in young males. But what explanation can be given for the fact that castrated males and females may also show aggressiveness, although its onset is later and its intensity reduced as compared to males? It is known that ovaries and the cortex of the adrenal glands also secrete some androgen, although the amount is considerably less than that produced by normally functioning testes.

Responsiveness to Androgenic Stimulation

Stimulation of the fetal brain by sex hormones plays a role in the masculinization process of mammals (Chapter 16), and the presence of androgens can accelerate the onset of aggressive behavior. Nevertheless, genetically produced differences in levels of aggressiveness or in male sexual activity can be largely caused by differences in physiological responsiveness to androgen stimulation. This relationship

Table 11-1   *Mean age of peck-right formation in intact cockerels, in untreated capons, and in capons injected with large doses of male sex hormone*

Group[a]	Treatment	Mean age at peck-right formation (weeks)
Cockerel	None	9.92
Capon	Androgen, from 10 days	7.82
Capon	Androgen, from 35 days	8.53
Capon	None	13.69

[a]Testes were surgically removed from males of the capon groups at 9 days.
SOURCE: Based on data presented by Guhl (1958).

has been shown in laboratory mammals (Valenstein et al., 1954; Riss et al., 1955) and in fowl (Hale, 1954; Ortman and Craig, 1968).

An example will demonstrate how genetic selection can modify sensitivity to androgens. Ortman and Craig (1968) selected for high and low social dominance ability within each of two breeds of chickens (White Leghorns and Rhode Island Reds). Strains within breeds differed considerably for aggressiveness as well as for dominance ability after five generations of selection. For the comparison of interest, males of the various strains were caponized before 2 weeks old and then kept until 4 months, when testing began. Males were then subdivided into groups and androgen was injected at different levels for different groups. Pair contests between males of the strains selected for high and low social dominance were conducted for capons receiving the same level of androgen (TP). The results, shown in Figure 11-1, are in terms of frequency of aggression (the dashed line) and as percentage of contests won by high line males (solid line) in those pairings in which agonistic behavior occurred. Two things are clearly evident: (1) androgen injections increase agonistic activity in males deprived of their own testicular source of male hormone, and (2)

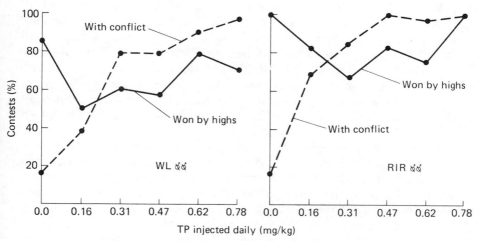

Figure 11-1 Results from pair contests between capons of strains selected for high and low social dominance ability within the White Leghorn (WL) and Rhode Island Red (RIR) breeds following testosterone propionate (TP) injections at five different levels. Percentages of 10-minute contests with aggression are shown by the dashed line and of contests won by strains selected for high social dominance by the solid line. (From Ortman and Craig, 1968.)

males of the strains that had become highly aggressive because of genetic selection were more likely to win pair contests when matched with low-strain males receiving the same amount of injected androgen. Changes produced by genetic selection were primarily caused by changes in physiological responsiveness rather than by differences in androgen secreted by the testes.

AVERSIVE STIMULI

Pain

Everyday observations indicate that animals attempt to push away or bite reflexively when pain is inflicted by a pinch or other minor injury. However, extremely painful or damaging conditions may lead instead to frantic escape behavior. O'Kelly and Steckel (1939) were among the first to observe that a shock delivered to the feet from an electric floor grid produced aggression in rats. Later studies have shown this to be a widespread phenomenon among animals, with the apparent exception of the guinea pig. Ulrich and Azrin (1962) tested different strains of rats for reflexive aggression when shocked; although all strains responded, differences were found in the intensity of shock required for comparable results. They found that fighting responses were much more likely to occur if the animals happened to be facing each other when the shock was delivered and also increased in frequency when they were forced to be close together in a small experimental chamber. The results suggest that the animals perceived their pain as having been caused by the other individual.

Other variables associated with pain and reflexive fighting have been reviewed (Ulrich, 1966). Age is an important influence; fighting between pairs of male rats increased dramatically in frequency when pairs were first shocked at ages varying from 24 (soon after weaning) until 93 days old. In castrated adult males, pain-induced fighting decreased over the next several weeks. Those experiments suggest that male sex hormone is in some way associated with reflexive aggression.

Brain Damage

Studies of controlling centers in the brain for aggression have progressed rapidly since the studies of Bard (1950), who demonstrated that removal of the outer layer (cortex) of the cerebrum of the cat produced animals with excessive defensive reactions. Such animals became highly irritable and would exhibit a "sham rage" accompanied by hair fluffing, spitting, and the other well-known characteristics of cats under attack. Those behaviors could be brought on in the operated cats by very minor kinds of stimulation, such as touching the cat. Continued high levels of irritability often persisted for months, long after the original surgical injuries had healed. Thus, the cerebral cortex as a whole appears to have an inhibitory effect on such aggressive behavior.

In contrast to the effect produced by removal of the entire cerebral cortex, removal of the neocortex or the amygdala portion only or damage to extensive regions of the hypothalamus produced cats that were either sleepy or docile and difficult to arouse to aggressive behavior when confronted with situations that would have aroused intact animals.

Brain Stimulation

The relatively crude approach of damaging or removing large portions of the brain, although initially helpful in gaining insight into general regions concerned with agonistic behavior, has now been replaced by the insertion of fine electrodes into precise locations with subsequent electrical stimulation and observations. Such methods have revealed the existence of numerous sites that control agonistic behavior. For example, Brown and Hunsperger (1963) reported that threat, attack, and escape behavior resulted from electrostimulation in areas of the amygdala, hypothalamus, and midbrain.

There appear to be two general types of response to brain stimulation which find outlets in aggressive behavior (Scott, 1971). The hypothalamus has areas that cause an emotional state similar to that of anger; a prolonged period of excitation results when those regions are stimulated. Other brain areas result in prompt responses of threat, aggression, or escape, which occur either simultaneously with stimulation or shortly after it stops.

Delgado and his associates (Plotnik et al., 1971) implanted electrodes into many brain sites in rhesus monkeys and then classified those as having positive, neutral, or negative properties when stimulated, as indicated by the monkey's responses in either pressing or lack of pressing a bar in order to obtain or prevent stimulation at each site. Each monkey was then fitted with a radio-controlled brain stimulator and tested in a social situation. None of the neutral or positive sites caused aggression when stimulated. However, painful foot shocks and stimulation of aversive brain sites had similar effects

Figure 11-2 Responses to aversive brain stimulation depend on social dominance rank. (a) Monkey A6 shows a submissive grimace following aversive brain stimulation in the presence of and without being threatened by its dominant partner. (b) Monkey A6 attacks a different partner, which is subordinate to it, following stimulation of the same brain area that caused the grimace in the presence of the dominant partner. (From Plotnik et al., 1971.)

on agonistic behavior. The results were especially interesting when aversively stimulated monkeys were paired with others known to be either submissive or dominant to them. When receiving a shock (to the brain *or* the feet), a dominant monkey would typically attack a subordinate. Conversely, when the stimulated monkey was itself subordinate, it would either exhibit no aggression, show submissive behavior, or redirect its aggression toward its own image in a mirror (Figure 11-2).

Frustration

Frustration of feeding behavior of hungry animals often causes aggressive behavior. An interesting observation is that a hungry pigeon trained to peck a key for continuous positive reinforcement (food being delivered each time) becomes agitated when the positive reinforcement is terminated. With food no longer delivered after each peck, the pigeon pecks rapidly at the key, and then if another pigeon is nearby will rush over and attack it about the head (Azrin et al., 1966).

An early study was carried out on the effects of competition for food in goats (Scott, 1948). Delayed feeding increased the amount of aggression shown, because subordinate animals would withstand more punishment in attempting to gain access to food when very hungry. Scott stated: "Yet there is no exception to the rule that it is only the dominant animals which become aggressive when frustrated."

Aggression caused by frustration of feeding behavior in chickens was studied by Duncan and Wood-Gush (1971). All comparisons were carried out in an experimental cage to which the birds had previously been accustomed. Food, when available, was presented in dishes reached through openings on opposite sides of the cage. Openings were only large enough for one bird to feed at a time. Hungry birds were frustrated by covering the feed bowl with clear plastic so that the food could be seen but not eaten. In all situations tested, socially dominant individuals showed increased aggression toward the subordinate bird when frustrated, and the frequency increased greatly when food was withheld for a full day as compared to shorter periods (Table 11-2).

Cockerels commonly exhibit "passive dominance" over hens. Thus, although they rarely pecked or threatened hens under ordinary circumstances, cockerels paired with hens were extremely abusive in the Duncan and Wood-Gush study when hungry and frustrated in their feeding attempts. Hens received 806 pecks and threats in a total of 8 hours of testing as compared to only 18 in a comparable time span when kept with nonfrustrated cockerels.

Table 11-2 *Relative frequency of aggressive acts by socially dominant members of pairs of chickens when hungry and frustrated in feeding attempts as compared to the same individuals when neither was hungry or frustrated*

	Test situation	
Pair composition	Length of food deprivation when frustrated (hr)	Frequency of aggressive acts when frustrated ÷ when not frustrated
Hens	2.5	1.2
	5.0	1.6
	7.5	2.6
	24.0	36.4
Cockerel and hen	24.0	44.8

SOURCE: Based on results presented by Duncan and Wood-Gush (1971).

Dominant individuals of many species become very aggressive when hungry and competing for food or when frustrated in their feeding attempts. This finding has led several investigators to use such techniques in working out dominance relationships. It is especially useful for pairs that do not otherwise interact frequently. Nevertheless, the results of a study by King (1965) indicate that the frustrating situation may become so extreme in some cases as to lead to questionable outcomes, particularly when levels of social tension are already high, as exist among young adult cockerels. King determined the peck order in each of three flocks of young males. He then compared frequency of aggression and the stability of dominance relationships by depriving flocks of their food supply for 24 hours and then presenting it for a 1-hour period in each of three ways. When food was spread evenly over the floor area, the frequency of aggression was very low, as all individuals were busy eating. When presented in a circular feeder allowing access by all individuals provided they crowded together, aggressive acts increased to 36 times that observed with the unrestricted area feeding and the incidence of subordinates attacking or threatening dominant penmates was 5%. When the feed was presented so that only one individual could feed at a time there was essentially an explosion of aggressive acts by the hungry cockerels and peck-order violations occurred at a frequency of about 50%. Dominance relationships quickly returned to their normal state in the absence of the extremely frustrating test situation.

What causes aggression when animals are frustrated? Several possibilities have been suggested, and more than one may be acting at the same time. Frustration is likely to be accompanied by excitement and moving about so that individuals encounter each other at higher frequency than usual; when animals are attracted to the same scarce resource, the personal space of dominant individuals may be invaded by subordinates; or the situation may be so aversive as to trigger agonistic behavior just as pain does.

LEARNING TO BE AGGRESSIVE

When aggression is rewarded, fighting may become a means of satisfying a need. Once initiated, such fighting may continue longer than required to obtain the reward, although final extinction of the fighting response is likely. These results were well illustrated by Ulrich et al. (1963), who succeeded in getting rats of a comparatively docile strain to fight vigorously by operant conditioning. They began by depriving the animals of water for 24 hours and then placing them in a chamber where one drop of water could be delivered at a time (accompanied by secondary reinforcers of a buzzer and light). After being trained to obtain water when the buzzer and light were activated, a control (nondeprived) rat was also placed in the cage at the same time as the thirsty one. Fighting behavior was then conditioned by making the buzzer, light, and water contingent upon approach and attack movements directed toward the control animal. Although no attacks occurred when first placed together, water-deprived rats' behavior was later shaped by this technique, so that controls were attacked and knocked about the cage.

After deprived rats were trained to attack control animals for drinks of water, they were then paired with each other. Vigorous fighting occurred and would often continue beyond the 2 to 3 seconds required for reinforcement. Even after the light and buzzer signaled that water was available, fighting would continue. Conditioned fighting was observed to be very similar to natural fighting behavior. It seems reasonable to speculate that part of the stimulus for continuing aggression beyond the time when water was made available was due to pain-induced reflexive fighting. When reinforcement was no longer available, there was a slight increase in fighting (caused by frustration?) and then extinction occurred.

Stimulation of "pleasure centers" of the hypothalamus can be used as a reward for aggressive behavior of rats toward other rats and also toward other species, such as cats and monkeys (Stachnic et al., 1966). Unlike stimulation to brain sites having aversive effects which elicit stereotyped, reflexive (unconditioned) responses, stimulation of pleasure centers is *not* typically associated with fighting. Nevertheless, pleasure-center stimulation can be used for eliciting fighting behavior by operant conditioning techniques in much the same way as food or water can be used with hungry or thirsty animals.

ISOLATION

It is well known in the pharmaceutical industry that keeping male mice in isolation after weaning increases their aggressiveness. Welch and Welch (1971) have shown that aggressiveness increases with the amount of time spent in isolation for periods up to 15 weeks. They kept male mice together in groups of 10 or singly and then placed pairs of strangers together which had been kept in the same way. The average number of fights per 5-minute test period was only two for animals kept in groups, but there were 9, 21, and 45 fights for those isolated for 2, 5, and 15 weeks.

CROWDING

Crowding usually occurs because the number of animals placed in a given space is increased. There are two components to crowding under this condition. If 100 individuals are placed in an area that was previously used for 25, two things have happened: group size has been quadrupled and area per animal has been reduced to one-fourth of what it was before. To examine the phenomenon fully, it should also be realized that there may be some sort of interaction involving both group size and space. Thus, the effect of decreasing space per animal by one-half may be very different with groups of four as compared to groups of 400. A few studies will be examined in which area per animal was varied while group size was held constant, and vice versa.

Invasion of Personal Space

Under some circumstances animals are attracted to each other and are very closely associated in space, as for mothers and their very young progeny, adult male and female during courting and mating, and when seeking shelter by means of other animals' bodies. However, it is also commonly observed, and the concept of personal space deals with this, that a minimum amount of space is maintained between animals (see Chapter 4). Dominant individuals usually reject the close approach of subordinates, and the latter may avoid the close presence of a more dominant animal even if no threat or aggressive act is given, probably because of earlier conditioning. It appears, therefore, that close proximity is aversive, except for special cases, as already noted.

   Chickens. A number of small studies with chickens at Kansas State University and some reports on laboratory mouse populations suggested a rather surprising relationship between area per individual and frequency of aggression; it appeared that as space decreased, the frequency of aggressive acts increased; then a sharp decrease resulted from further crowding (Polley et al., 1974). Those results were checked by a carefully designed experiment in which the number of hens was held constant as space per bird was varied (Al-Rawi and Craig, 1975). Effects of varying space per hen on frequency of aggression are shown in Figure 11-3.

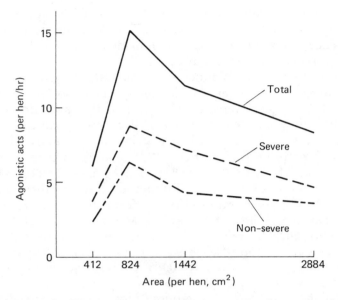

Figure 11-3 Effect of area per hen, in groups of constant size (four hens per flock), on total, severe, and nonsevere agonistic acts hourly. With intermediate amounts of space highest frequencies of aggression occur; it has been suggested that with large areas "personal space" is violated less, but with extremely high density "third-party" effects become important. (From Al-Rawi and Craig, 1975.)

As in the earlier study, Al-Rawi and Craig found that an intermediate amount of space per individual was associated with the highest frequency of agonistic acts among hens. How can such results be explained? It seems likely that with large amounts of space, hens can move about rather freely without violating the personal space of others. As space decreases, hens are forced together and dominant ones are more likely to peck or threaten subordinates. But what happens with very little space? Suddenly, the frequency of aggression drops off drastically.

Edinburgh researchers Hughes and Wood-Gush (1977) also found that under very high density conditions (multiple-hen cages), there was reduced aggressive behavior as compared to intermediate-density floor pens. Their observations of hens in floor pens suggested that normal threat displays require minimal amounts of space exceeding the dimensions of most cages. Indeed, the incidence of threats is clearly reduced in cages (Bhagwat and Craig, 1979), but the reduction of aggressive head pecking found in high-density environments must be explained otherwise. Hughes and Wood-Gush suggested that with extreme crowding subordinate hens may not trigger pecking by a dominant bird if they are already within the dominant's sphere of influence; only entry into an individual's personal space was postulated as causing such behavior.

Results obtained by Ylander and Craig (1980) support a hypothesis advanced earlier (Craig et al., 1969) that a socially dominant "third party" inhibits interactions between pairs that would otherwise occur when they are close together. Males were found to be particularly effective inhibitors of aggressive behavior by dominant members of pairs of feeding hens. With a cock in their immediate presence, pairs of hens had only 5 peck-avoidances in 24 ten-minute trials, when the cock was at a distance of 1 meter there were 21, and when he was temporarily removed from the pen there were 74. Threat-avoidances followed the same pattern, but the effects were less pronounced. The same basic experiment was repeated, except that the hen at the top of each small flock's peck order was substituted in place of the male. Roughly comparable results were obtained as shown in Table 11-3. However, the inhibitory effect on peck avoidances appeared to be unquestionably present only in the dominant hen's close presence.

When pairs of cockerels were kept in roomy, solid-floored cages from 12 until 20 weeks of age, they could be classified easily into dominant and subordinate categories by 20 weeks because the subordinates showed clear signs of physical abuse and submissive posture, providing indirect evidence of frequent and severe aggression by the dominant member (Grosse and Craig, 1960). Subordinate males were also delayed in attainment of sexual maturity. In a later study (Craig and Polley, 1977) pairs of cockerels were kept in much smaller, wire-floored cages, allowing only 30% as much space per cockerel. Over an experimental period of about 24 weeks, the pairs

Table 11-3 *Dominant hen effect on frequency of peck-avoidances between lower-status hens*

| Flock | Distance (or absence) of dominant hen from feeding pair | | | Probability that differences are due to chance only |
	0 m	1 m	Absent	
1	0[a]	17	5	<0.01
2	0	3	12	<0.01
3	7	17	25	<0.05

[a]Total acts for ten 10-minute trials.

SOURCE: From Ylander and Craig (1980).

that were extremely crowded had essentially equal weight gains and survival as compared to singly caged males. Cockerels kept as pairs in the second study showed no signs of physical abuse inflicted by one individual on the other (although their feathers were worn from rubbing on the wire cages); perhaps they were so crowded that space was not available for fighting.

Dairy Cattle. Signoret and Bouissou (1971) reported, from an earlier study of Bouissou, on the frequency of aggression for 10 heifers when kept on pasture and in open housing at two densities. Square meters per heifer of 5500 (on pasture), 12, and 6 were associated with 2.2, 9.0, and 22.6 aggressive interactions per animal per hour, respectively; frequencies of aggression increased as space per animal decreased. (More than half of the aggressive interactions on pasture included physical contact as compared to about 17% of all acts in confinement, but even so the actual frequency of "blows" per hour was considerably less on pasture.) Space allowances for dairy cows in confinement systems of management in the United States were believed to average about 8 square meters per cow in 1974, according to Arave et al. (1974). They carried out a short-term study to determine what effect sharp reductions in space (from 9.3 to 2.3 square meters per cow) might have in a group of 17 Holsteins. Ample free-stall and feed-bunk space were provided. Cows in the restricted-space phases of the study (weeks two and four of a 4-week comparison) walked about less, had fewer social encounters, fewer leukocytes in their milk (high counts are taken as an indication of stress), and did not differ in milk yield. Arave et al. concluded: "In terms of herd management, restricting cows to a lot size of 2.3 m^2/cow was actually beneficial." Longer-term studies are obviously desirable.

Swine. Ewbank and Bryant (1972) were prompted to test space allowances for pigs by the British Code of Recommendation for growing pigs (M.A.F.F., 1971), which requires 0.74 m^2 (about 8 ft^2) per animal to 90 kg (about 200 lb) of body weight. They set up three pen sizes so that pigs would have approximately 0.56, 0.77, or 1.19 m^2 (6, 8.3, or 12.8 ft^2) of floor space each. Group size was held constant at eight, with four barrows and four gilts per pen. Linear dominance orders were generally found. Dominant animals usually reinforced their status by aiming blows at the head, neck, or shoulder. Subordinates often showed "token" retaliation. Agonistic interactions were recorded and the "activity complex" in which they occurred. Thus, the frequency of total agonistic acts could be determined for pigs while both were feeding, rooting, and so on, as shown in Table 11-4. Their results confirm what is generally known; animals are more likely to behave aggressively during feeding than when engaged in other activities. Although feeding occurred only 30% of the time while pigs were observed, 60% of all agonistic acts were observed while the pigs were feeding.

The effect of decreasing space on agonistic activity is seen most clearly in the Ewbank and Bryant study by comparing the increase occurring with the smallest as compared to the largest area allowance per pig, as shown in Table 11-5. Not only did the relative frequency

Table 11-4   *Agonistic interactions and activity complex of swine in dry lots*

Activity complex	Percent of total agonistic acts	Comments
Both feeding	60	Interactions tended to be "intense" during feeding
Both standing	19	
Dominant resting, subordinate standing	14	Usually occurred when a dominant pig was disturbed
Both resting	4	
Dominant standing, subordinate resting	2	
Both rooting	1	
	100	

SOURCE: Based on results of Ewbank and Bryant (1972).

Table 11-5   *Relative frequency of agonistic acts among swine at two densities*

Activity complex and kind of agonistic activity	Percent increase at highest over lowest density
Pigs feeding	
Agonistic acts (total)	10
Open-mouth attacks (most severe)	70
Dominance-order violations	100
Pigs standing	
Agonistic acts (total)	150
Open-mouth attacks	30
Resting pig attacks standing pig	
Agonistic acts (total)	115
Open-mouth attacks	50

SOURCE: Calculated from data presented by Ewbank and Bryant (1972).

of agonistic acts increase with less space per pig, but there was a higher percentage of the most severe and potentially damaging "open-mouth" attacks. It was concluded that "Failure to allow sufficient space appears to lead to a breakdown in communications, with a resulting increase in social strife."

General. Experimental results with cows and hens suggest that aggression increases as space per animal decreases, at low to moderate densities, but aggression decreases precipitously as density reaches very high levels. Would pigs, kept on slotted floors at even higher density than tested by Ewbank and Bryant (1972), show decreased frequencies of agonistic acts? Although extreme crowding may reduce overt aggression, its effects on well-being and productivity must be considered separately. It will be seen later that frequency of aggression, in itself, is frequently an inadequate index of behavioral stress.

GROUP SIZE

How many individuals can be remembered? Each person knows his own limitations in remembering names and recognizing faces. Guhl (1953) presented partial evidence that a complete peck order existed in a flock of 96 hens, and Albright (1978), on the basis of limited evidence, suggested that the upper limit for memory of individuals may be about 100 in cattle. In a later study with chickens, Craig and Guhl (1969) observed samples of hens in flocks ranging in size from 100 to 400. They found that hens in larger flocks (200 or more) tended to stay in particular areas and were more dominant where they spent more time. Remaining in particular areas would limit the need to recognize more hens. Thus social chaos associated with meeting many strangers may be prevented in larger flocks by the tendency of hens to stay in their own neighborhood.

Even though flocks of 100 or more may be socially organized, Banks and Allee (1957) found that "peck-order violations" became more frequent as group size increased from 6 to 24. A peck-order violation consists of a subordinate individual behaving aggressively toward its social superior. Banks and Allee set up several small flocks of hens in spacious floor pens and observed them closely for periods of 4 to 5 months. After peck orders were established, it was noted that low-status hens were occasionally pecking higher-ranking ones; the relative frequency was associated with group size as shown:

Flock size	Average number of violations per hen in 18 hours of observation
6	4.8
12	13.7
24	20.3

It is clear that more violations occurred in the larger flocks. Such violations were ineffective, as insubordination was generally punished promptly and no reversals of rank occurred. It was suggested that reinforcement of status for particular hens was less frequent in larger groups and that limits of recognition were being approached. Evidence from other studies indicates that peck orders are relatively stable in even larger flocks than studied here, yet temporary confusion and lack of recognition may be responsible for higher levels of aggression in larger flocks.

Agonistic behavior has been closely observed in multiple-hen cages. The frequency of aggressive acts per hen increases as group size increases from 4 to 28 (Al-Rawi and Craig, 1975). Most aggressive acts in colony cages are pecks rather than threats and most occur while the birds are feeding or near the feeder. In a second study (Al-Rawi et al., 1976) involving cages with flocks of size 4, 8, and 14, higher levels of aggression were observed in the larger flocks during the first 8 weeks, but agonistic activity was much reduced for all flocks when they were observed again, 26 weeks after assembly. Perhaps hens forced to live in the close proximity of multiple-hen cages become so well acquainted that temporary failure of recognition does not occur, even in flocks of 8 and 14.

Cattle in herds of several thousand may be organized into subgroups, and individuals may simultaneously be a member of two or three different dominance orders (Hafez and Bouissou, 1975). Hulet et al. (1975) commented on the continued "baaing" in large commercial flocks of sheep in contrast to small naturally formed flocks, which are much quieter, and imply that these auditory signals result from family members attempting to locate and stay with each other. It would appear beneficial for animals to remain in groups where individuals know each other to prevent excessive agonistic activity.

AGGRESSION AMONG STRANGERS

When adolescent or mature animals or birds that are strangers to each other are placed together in an unfamiliar place, there is usually a brief lull in activity during which they survey their new surroundings and companions. Then one or more pairs begin interacting agonistically. Males may engage in vicious battles; females usually interact less vigorously, as in most breeds of chickens, pigs, cattle, and horses, or may show little activity, as in sheep and goats. Breed and strain differences are likely to be noted in intensity and duration of contests for dominance status. Cocks and dogs of "game" or fighting stock may engage in mortal combat unless separated, but in most groups the outcome is rather quickly decided within pairs. If the assembled groups are relatively small, the dominance status of all possible pairs may be tentatively established within a few hours, although agonistic interactions may continue for considerably longer periods as dominant animals continue to reinforce their newly acquired "right" of aggressive behavior toward subordinates without fear of effective retaliation.

Why do strange animals attempt to establish their dominance relationships promptly? They usually exhibit aggressive behavior toward strangers, at least toward others of roughly comparable size, even when not competing for any obvious necessity other than personal space. It is unlikely that a single explanation will suffice for all cases, but it seems likely that past experience is involved; most adolescent or mature animals will have been rewarded in past competitive situations by being socially dominant and are conditioned therefore to seek it. Passive acceptance of subordinate status in the presence of an overwhelmingly large or obviously powerful or combative individual is also likely to have been learned as a result of negative reinforcement in the past.

INTERNAL DRIVE FOR AGGRESSION

Is there an internal drive for aggressive behavior? If such a drive exists, is it cyclical, as hunger is? Scott (1971) argues in opposition to the view that "aggression-specific energy" arises spontaneously in the nervous system and accumulates until discharged, as expected according to the "hydraulic theory" (Lorenz, 1950). Instead, Scott proposes that anger and fear are the emotions primarily responsible for social fighting and that those are aroused by external stimuli. Those emotions, once aroused, may prolong and magnify reactivity to external stimuli, particularly if behavior is blocked. (Consider the effects of feeding frustration.) Scott further proposes that in the absence of further external stimulation, the internal arousal associated with anger and fear die out. Clearly, these are two very different theories as to basic causation of social fighting.

Aggression among Agonistically Naive Animals

It has been suggested that aggression arises among young animals because play occurs spontaneously and becomes more vigorous as age increases until pain is inadvertently inflicted on one animal by another. It is known that pain elicits defensive or reflexive fighting behavior. If such an unconditioned aggressive response produces a reward (the other animal withdraws), then aggressive behavior is reinforced, at least toward individuals that avoid or stop inflicting pain. What if a fight should break out when one animal inflicts pain on the other and retaliation occurs? Then a dominance relationship may be learned; the individual that dominates is rewarded (as before) and the one that becomes subordinate is rewarded by submissive behavior, as it is no longer attacked.

Animals that are reared together do not make suitable subjects to determine whether there is an internal drive for aggressive behavior because of the learning process and difficulty of the observer in knowing when rough play stops and attacks with aggressive intent begin. Some species, such as sheep and dogs, have such powerful affiliative drive that they become greatly disturbed and are stressed if deprived of companions. On the other hand, chickens and calves may be reared in isolation without obvious stress until those ages at which agonistic behavior occurs in undisturbed groups.

Male chicks were isolated visually and physically from each other within 24 hours of hatching and maintained in that way until 10 weeks of age by Ratner (1965). He observed the development of "social" behaviors closely and compared them with those of group-reared chicks kept as controls. Such activities as aggressive pecks, threats, and fights could not be recorded for isolated chicks, but movements typical of "sparring" and juvenile fights were observed. There was a delay in onset; sparring motions occurred at about 14 and juvenile fighting movements at 36 days for isolated males as compared to 10 and 24 days for group-reared ones. When Ratner assembled the isolated males at 10 weeks (slightly beyond the normal age for peck-order formation), all males showed nearly immediate and intense agonistic behavior, and full social organization was present within several days. The initial aggressive acts were poorly directed, pecks being delivered to the backs, wings, and heads of the others. It was estimated that 200 to 250 pecks were exchanged within 2 hours of assembling 14 males. By the day after assembly, aggressive pecks were being delivered to the head exclusively.

Ratner's results with chickens suggest that an internal drive for aggressive behavior may exist in that species (although he did not explicitly state such a conclusion). The literature is remarkably deficient in clear-cut evidence bearing on this question for other animals. Certainly, we know that breeds of cattle, dogs, and chickens have been successfully selected for easy arousal of aggression and fighting ability. Other breeds are known for their social tolerance and peaceful dispositions. Such differences indicate that genetic factors can predispose animals to be more or less aggressive.

Is Aggression Cyclic?

It is important to know whether aggressive behavior, if instinctive in nature, is also cyclic. It seems clear that aggression can be rewarding when it serves as a "means to an end," but to answer the basic question it is necessary to know whether the likelihood of aggression increases with the time elapsed since the last aggressive interaction. At present, the answer is unknown.

REFERENCES

ALBRIGHT, J. L., 1978. Social considerations in grouping cows. *In* C. J. Wilcox (Ed.), Large Dairy Herd Management. University Presses of Florida, Gainesville, Fla.

AL-RAWI, B., and J. V. CRAIG, 1975. Agonistic behavior of caged chickens related to group size and area per bird. Appl. Anim. Ethol. 2:69-80.

——, and A. W. ADAMS, 1976. Agonistic behavior and egg production of caged layers: genetic strain and group-size effects. Poult. Sci. 55:796-807.

ARAVE, C. W., J. L. ALBRIGHT, and C. L. SINCLAIR, 1974. Behavior, milk yield and leucocytes of dairy cows in reduced space and isolation. J. Dairy Sci. 57:1497-1501.

AZRIN, N. H., R. R. HUTCHINSON, and D. F. HAKE, 1966. Extinction induced aggression. J. Exp. Anim. Behav. 9:191-204.

BANKS, E. M., and W. C. ALLEE, 1957. Some relations between flock size and agonistic behavior in domestic hens. Physiol. Zool. 20:255-268.

BARD, P., 1950. Central nervous mechanism for the expression of anger in animals. *In* M. L. Reymart (Ed.), Feelings and Emotions. McGraw-Hill Book Company, New York.

BHAGWAT, A. L., and J. V. CRAIG, 1979. Effects of male presence on agonistic behavior and productivity of White Leghorn hens. Appl. Anim. Ethol. 5:267-282.

BOUISSOU, M. F., and S. ANDRIEU, 1977. Etablissement des relations de dominance-soumission chez les bovins domestiques. IV. Etablissement des relations chez les jeunes. Biol. Behav. 2:97-107.

——, 1978. Etablissement des relations préférentielles chez les bovines domestiques. Behaviour 64:148-157.

BROWN, J. L., and R. W. HUNSPERGER, 1963. Neurothology and the motivation of agonistic behaviour. Anim. Behav. 11:439-448.

CRAIG, J. V., and A. M. GUHL, 1969. Territorial behavior and social interactions of pullets kept in large flocks. Poult. Sci. 48:1622-1628.

CRAIG, J. V., and C. R. POLLEY, 1977. Crowding cockerels in cages: effects on weight gain, mortality and subsequent fertility. Poult. Sci. 56:117-120.

CRAIG, J. V., D. K. BISWAS, and A. M. GUHL, 1969. Agonistic behavior influenced by strangeness, crowding and heredity in female domestic fowl. Anim. Behav. 17:498-506.

DUNCAN, I. J. H., and D. G. M. WOOD-GUSH, 1971. Frustration and aggression in the domestic fowl. Anim. Behav. 19:500-504.

EWBANK, R., 1967. Behavior of twin cattle. J. Dairy Sci. 50:1510-1512.

——, and M. J. BRYANT, 1972. Aggressive behaviour amongst groups of domesticated pigs kept at various stocking rates. Anim. Behav. 20:21-28.

GROSSE, A. E., and J. V. CRAIG, 1960. Sexual maturity of males representing twelve strains of six breeds of chickens. Poult. Sci. 39:164-172.

GUHL, A. M., 1953. Social behavior of domestic fowl. Kans. Agric. Exp. Sta. Bull. 73.

——, 1958. The development of social organisation in the domestic chick. Anim. Behav. 6:92-111.

HAFEZ, E. S. E., and M. F. BOUISSOU, 1975. The behaviour of cattle. *In* E. S. E. Hafez (Ed.), The Behaviour of Domestic Animals, 3rd ed. The Williams & Wilkins Company, Baltimore.

HALE, E. B., 1954. Androgen levels and breed differences in the fighting behavior of cocks. Bull. Ecol. Soc. Am. 35:71-72.

HUGHES, B. O., and D. G. M. WOOD-GUSH, 1977. Agonistic behaviour in domestic hens: the influence of housing method and group size. Anim. Behav. 25:1056-1062.

HULET, C. V., G. ALEXANDER, and E. S. E. HAFEZ, 1975. The behaviour of sheep. *In* E. S. E. Hafez (Ed.), The Behaviour of Domestic Animals, 3rd ed. The Williams & Wilkins Company, Baltimore.

KING, M. G., 1965. Disruptions in the pecking order of cockerels concomitant with degrees of accessibility to feed. Anim. Behav. 13:504-506.

LORENZ, K. Z., 1950. The comparative method in studying innate behaviour patterns. Symp. Soc. Exp. Biol. 4:221-269.

M.A.F.F., 1971. Codes of recommendations for the welfare of livestock. Code No. 1, Cattle. Code No. 2, Pigs. Code No. 3, Domestic fowls. Code No. 4, Turkeys. Code No. 5, Sheep. Ministry of Agriculture, Fisheries and Food, London.

O'KELLY, L. W., and L. C. STECKLE, 1939. A note on long-enduring emotional responses in the rat. J. Psychol. 8:125-131.

ORTMAN, L. L., and J. V. CRAIG, 1968. Social dominance in chickens modified by genetic selection–physiological mechanisms. Anim. Behav. 16:33-37.

PLOTNIK, R., D. MIR, and J. M. R. DELGADO, 1971. Aggression, noxiousness, and brain stimulation in unrestrained Rhesus monkeys. *In* B. E. Eleftheriou and J. P. Scott (Eds.), The Physiology of Aggression and Defeat. Plenum Press, New York.

POLLEY, C. R., J. V. CRAIG, and A. L. BHAGWAT, 1974. Crowding and agonistic behavior: a curvilinear relationship? Poult. Sci. 53:1621-1623.

RATNER, S. C., 1965. Comparisons between behaviour development of normal and isolated domestic fowl. Anim. Behav. 13:497-503.

RISS, W., E. S. VALENSTEIN, J. SINKS, and W. C. YOUNG, 1955. Development of sexual behavior in male guinea pigs from genetically different stocks under controlled conditions of androgen treatment and caging. Endocrinology 57:139-146.

SCHAKE, L. M., and J. K. RIGGS, 1970. Activities of beef calves reared in confinement. J. Anim. Sci. 31:414-416.

SCHLOETH, R., 1961. Das Sozialleben des Camargue-rindes. Qualitative und quantitative Untersuchungen über die sozialen Beziehungen—insbesondere die soziale Rangordnung—des halbwilden französischen Kampfrindes. Z. Tierpsychol. 18:574-627.

SCOTT, J. P., 1948. Dominance and the frustration-aggression hypothesis. Physiol. Zool. 21:31-39.

——, 1971. Theoretical issues concerning the origin and causes of fighting. *In* B. E. Eleftheriou and J. P. Scott (Eds.), The Physiology of Aggression and Defeat. Plenum Press, New York.

SIGNORET, J. P., and M. F. BOUISSOU, 1971. Adaptation de l'animal aux grandes unités. Etudes de comportement. Bulletin Technique d'Information 258:367-372. Ministère de l'Agriculture, France.

STACHNIK, T. J., R. ULRICH, and J. H. MABRY, 1966. Reinforcement of intra- and inter-species aggression with intracranial stimulation. Am. Zool. 6:663-668.

STEPHENS, D. B., 1974. Studies on the effect of social environment on the behaviour and growth rates of artificially-reared British Friesian male calves. Anim. Prod. 18:23-34.

ULRICH, R., 1966. Pain as a cause of aggression. Am. Zool. 6:643-662.

——, and N. H. AZRIN, 1962. Reflexive fighting in response to aversive stimulation. J. Exp. Anal. Behav. 5:511-520.

ULRICH, R., M. JOHNSTON, J. RICHARDSON, and P. C. WOLFF, 1963. The operant conditioning of fighting behavior in rats. Psychol. Rec. 13:465-470.

VALENSTEIN, E. S., W. RISS, and W. C. YOUNG, 1954. Sex drive in genetically heterogeneous and highly inbred strains of male guinea pigs. J. Comp. Physiol. Psychol. 47:162-165.

WELCH, A. S., and B. L. WELCH, 1971. Isolation, reactivity and aggression: evidence for an involvement of brain catecholamines and serotonin. *In* B. E. Eleftheriou and J. P. Scott (Eds.), The Physiology of Aggression and Defeat. Plenum Press, New York.

YLANDER, D. M., and J. V. CRAIG, 1980. Inhibition of agonistic acts between domestic hens by a dominant third party. Appl. Anim. Ethol. 6:63-69.

12

Status

Status refers to position or rank in relation to others. It may concern an individual of a pair in which case the individual is usually classified as being either dominant or subordinate, or less commonly, each member of the pair may be of equal status. Status may also refer to the position of an individual or of a particular group in a hierarchy and be expressed in terms of rank, dominance value, or a similar measure. Social hierarchies are frequently referred to as *peck orders*, although this terminology is obviously most appropriate when applied to organized groups of chickens or birds in which pecking is a principal means of reinforcing status by dominant birds. Terms such as peck order, dominance order, and social hierarchy will be used interchangeably in what follows.

The meaning of the words "dominant" and "aggressive" as well as "subordinate" and "submissive" need to be clearly differentiated. *Aggressive* behavior involves some form of attack or threat, but dominant individuals range in aggressiveness from complete tolerance (never attacking or threatening) to nearly constant abuse of subordinates. Although *dominance* is usually attained initially by some form of aggression, it need not be. Some individuals that are very large or show great confidence by their posture or movement may achieve dominant status because of avoidance by others without showing any overt aggressive behavior. Thus, an imposing cock or bull may often be placed with a group of females or with younger and smaller individuals with nearly immediate passive acceptance of *subordinate* status by the others. In a similar way, subordinates may not show obvious *submissive* behavior in the presence of a dominant, but tolerant individual, whereas they may show extreme and active avoidance or appeasement behavior in the presence of an abusive dominant.

Contests involving dominance status are of considerable importance to the individual in group-living species, where high status provides ready access to food, mates, preferred locations, and other desirable conditions that influence well-being and biological fitness. However, excessive aggressiveness may be maladaptive to both the individual and the group, particularly in environments provided by man, where care is taken to provide all essential requirements. The group as a whole benefits and is more productive if dominance relationships are quickly established and energy is not dissipated by continuing aggressive behavior. Excessively abusive individuals may become less productive and have lowered fitness because of unnecessary activity, and in animals kept for reproduction, by reduction in sexual activity.

Submissive behavior has considerable adaptive value for smaller, weaker, and younger animals. Such behavior allows subordinate individuals to coexist with more aggressive dominant ones. Temporary escape or appeasement displays permit subordinates to benefit from being allowed to remain with the group. Appeasement displays tend to reduce or inhibit further aggression by either making the animal appear less threatening (by lowering the head, crouching, turning away, "freezing," or similar acts) or by distracting the aggressor with behavior more characteristic of juveniles or sexually receptive females.

SCHJELDERUP-EBBE AND PECK ORDERS

It has been known from ancient times that some animals are socially dominant to others. Even so, a brilliant and charming study of domestic chickens by the Norwegian Schjelderup-Ebbe, published in 1922, revealed to social psychologists of the time something that apparently surprised them: animal groups could be highly structured.[1] Although flavored by an anthropomorphic viewpoint, most of his observations and conclusions have been verified. They have given profound insight into the workings of chicken society, in particular, and into many other animal societies as well. Those insights have been most helpful in understanding group-living, domesticated animals.

---

[1] The classical paper of Schjelderup-Ebbe (1922a) has been translated from "Norwegian German" to "American English" (1922b) by the combined efforts of M. Schleidt, W. M. Schleidt, and M. W. Schein, as indicated in the reference section.

Schjelderup-Ebbe described the everyday existence of chickens on the farm as follows (translated):

> Anyone who thinks the inhabitants of a chicken yard are thoughtless, happy creatures with a daily life of undisturbed pleasure, at peace with each other and, without a worry about anything, crow, lay eggs and eat, is thoroughly mistaken. A grave seriousness lies over the chicken yard and hens exhibit much anger and fear. Even the liveliness demonstrated during feeding time is not in fact happy liveliness, since it is all based on the effort of each hen to consume as much as possible herself and to drive away the others as much as possible.

Elsewhere, he gives a clue as to how peck orders may be worked out rapidly: "If one throws feed to the hens, they come running from all sides in order to ingest the greatest possible amount. . . . There is a constant quarrel over the food; each hen shows clearly that she wants to be first to eat. . . ." Peck orders were described with the observation that pecking is undirectional among well-acquainted hens; thus, "if hens A and B live together for some time and hen A pecks at hen B, then B will not peck at A; instead B is afraid of A and avoids her."

Schjelderup-Ebbe pointed out that peck orders, especially those involving more than a few birds, are likely not to be "continuous," or as we would probably say, linear. Pecking triangles, squares, and other complex social structures were recognized. In a pecking triangle each bird pecks one of the others and is pecked by the third, so that no one individual is either dominant or subordinate to the other two. This may be shown diagrammatically for individuals A, B, and C as

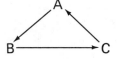

Arrowheads indicate the direction of pecking; here A pecks B, B pecks C, and C pecks A. In diagramming a linear peck order it is customary to show a simplified form, thus:

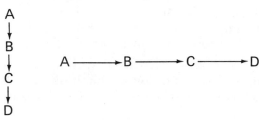

Here it is understood that A pecks not only B, but also C and D [i.e., anything below (or to the right) is subordinate]. B is then known to be dominant to both C and D, but is subordinate to A; C pecks D only; and D is subordinate to A, B, and C.

Among Schjelderup-Ebbe's findings, confirmed by others, are the following:

1. Aggressiveness is most evident when individuals actively compete for something, as when grain is placed before hungry hens or when nesting or roosting places are sought. However, a lower frequency of pecking is also observed even when individuals are not obviously competing.

2. Social structure of groups may be simple or complex (see above), but each individual knows its status relative to every other one (i.e., it has either a dominant or subordinate position relative to every other individual).

3. Pair relationships among females are relatively stable, often persisting for years. Although peck-order violations may occur, reversals of dominance are uncommon, and if occurring at all, are likely to follow a hard fight.

4. Low-status individuals often get little feed, are chased about, and appear to be nervous and frightened.

5. When strange adults meet, they are likely to establish a dominance relationship promptly.

6. Some larger and obviously stronger individuals may be observed being dominated by smaller ones. Such relationships may be caused by such experiences as these:

   a. Older individuals establish dominance easily over younger and smaller ones, and the relationship continues even after the younger animal becomes larger and stronger.

   b. Strangers introduced into an organized flock are usually dominated; perhaps because they are frightened when introduced.

   c. One hen may be attacked by more than one other individual simultaneously and thereafter becomes subordinate to each of those which attacked together.

7. Physiological changes associated with such events as coming into reproductive condition or moulting can influence aggressiveness and occasionally result in changes in social status.

Schjelderup-Ebbe commented briefly, that "pecking in a flock of roosters is the same as the pecking in a flock of hens, with the exception that the roosters are fiercer." Masure and Allee (1934), following leads of Schjelderup-Ebbe, compared single all-male and all-female flocks and found that the male flock had a more complex structure (more pecking triangles), some dominance relationships took a longer time to develop, and the peck order was less stable.

DETERMINATION OF STATUS

Competitive Feeding

The use of competitive feeding trials for establishing dominance relationships has been widely used. Even when feed was continuously available, significant positive associations were found between within-flock social ranks and number of times feeding for hens (Tindell and Craig, 1959); higher-status individuals spent more time at the feeder. It was shown in Chapter 11 that feeding frustration and limited access to feed increases frequency of aggressive activity, and the direction of aggression is generally the same as observed under "normal" conditions. However, King (1965) found that with very hungry cockerels and feed available for only one individual at a time, there was a high incidence of peck-order violations, which would have confused an observer trying to determine dominance relationships. Lowry and Abplanalp (1970, 1972) were able to determine dominance relationships rapidly for pairs of hens in cages and for small flocks in floor pens by making hens moderately hungry and then providing feed so that only one of a pair or not all of a group could feed at once.

A particularly useful study was carried out by Sereni and Bouissou (1978) with saddle horses. Social organization was determined by extensive observations of three groups of animals while they were grazing on pasture. Dominance relations determined in that way were compared with those obtained from very brief competitive feeding trials. Groups on pasture were observed for 30, 50, or 200 hours, and all agonistic acts, including different kinds of threats, bites, kicks, charges, and avoidances, were recorded. Subordinate horses sometimes showed aggressive behavior while on pasture, but Sereni and Bouissou considered the one showing more aggressive behavior within the pair as the dominant member.

Sociograms based on the behavior of the horses while on pasture are shown in Figure 12-1. Within the three groups there was a total

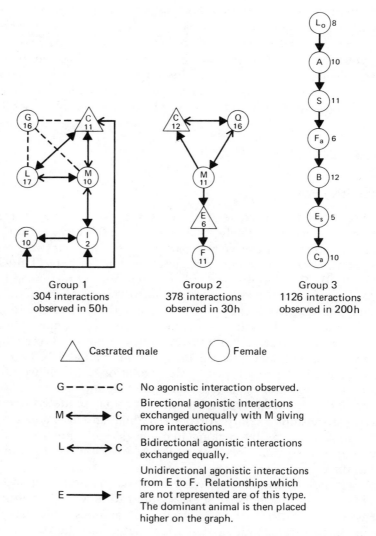

Group 1
304 interactions
observed in 50h

Group 2
378 interactions
observed in 30h

Group 3
1126 interactions
observed in 200h

△ Castrated male   ◯ Female

G– – – –C   No agonistic interaction observed.

M◀———▶C   Birectional agonistic interactions exchanged unequally with M giving more interactions.

L◀———▶C   Bidirectional agonistic interactions exchanged equally.

E———▶F   Unidirectional agonistic interactions from E to F. Relationships which are not represented are of this type. The dominant animal is then placed higher on the graph.

Figure 12-1 Sociograms of groups of horses. The animals are represented by the first letter of their name. The number indicates their age in years. (From Sereni and Bouissou, 1978.)

of 46 possible dominance relationships [when group size is n, there are $(n)(n-1)/2$ dominance relationships]. Dominance was established in 41 of the 46 possible pairs. Group 3 had a linear dominance order, group 2 included a triangular relationship, and group 1 had five undetermined pair relationships. In group 1, L-C and F-I had equal frequencies of aggressive acts, and G was not observed to interact with C, L, or M. G frequently remained apart from the others.

When food competition tests were carried out with the same animals (by Sereni and Bouissou), two animals were released at a time into a paddock after having fasted in a stable overnight. A bucket containing oats was placed in the paddock and the test period, lasting either 2 or 3 minutes, began as soon as one began to feed. Only one horse could feed at a time, and the amount of time each controlled the bucket was timed. Control time was defined as that time when one animal prevented the other from feeding either because its head was inside the bucket or because its presence was sufficient to keep the other from feeding. All possible pairs within each group were tested in this way, and each pairing was repeated either two or three times. Aggressive acts were of the same type as seen on pasture. Almost without exception, interactions while competing for feed were unidirectional.

In comparing their observations, made while horses were on pasture and in feeding competition trials, Sereni and Bouissou found that dominants controlled the feed bucket for average times varying from 136 to 165 out of 180 seconds for those pairs where the dominance relation was already known from pasture observations. Group 1, which had five undecided pair relationships after 50 hours of observation on pasture, presented much less of a problem in the competitive feeding tests; dominance, as indicated by priority of access to feed, remained undecided in only one pair. Average feeding time of the dominant animal in those four pairs that were determined by feeding trials only was 160 out of 180 seconds. The undecided pair (I–F) tended to share the feed and had no aggressive interactions.

Sereni and Bouissou concluded that competitive food tests with horses give very clear results in nearly all cases and that those results were consistent with those from pasture observations where the dominance situation was known. Tyler (1972) also found nearly perfect agreement between dominance relationships of horses under unrestricted conditions and priority of access to hay. This type of test allows relatively prompt determination of dominance relationships within pairs of acquainted animals.

A very similar food competition test to that of Sereni and Bouissou was devised and used by Houpt et al. (1978) in 11 herds varying in size from 4 to 11 horses. Adult animals rarely shared food during the tests, but it was not uncommon among lower ranking colts and fillies. Feeding trials lasted five times as long in the study of Houpt et al. as compared with that of Sereni and Bouissou, so that competition may have been less during the latter part of longer feed trials. Dominance hierarchies based on priority of access to feed by the hungry horses indicated that linear hierarchies were prevalent within small groups, but triangular relationships were common within larger ones.

Limitations of Competitive Feeding Trials

Feeding trials have commonly been carried out between animals that know each other and in an environment familiar to them. When strangers are to be compared for aggressiveness or dominance ability, they are usually placed together in a "neutral" area that is strange to both. Food placed before hungry individuals in such a situation may be ignored, at least in tests of relatively limited duration. Thus, Al-Rawi and Craig (1976) found no difference in frequency of aggression or of dominance decisions for chickens placed together in 10-minute initial pair contests, depending on the presence or absence of a feeding stimulus. Although hungry, the birds ignored the feed; moving the contestants to the pair-contest pen and the presence of a stranger were sufficiently distracting to offset the feeding stimulus.

Some criticisms have been directed at the use of food-competitive situations for establishing dominance relationships, particularly for ungulates, as the localization of a food source is rare under natural conditions. Nevertheless, under domestication, localized distribution of food is common, and as Houpt et al. (1978) state in the case of horses: "There is no question that equine dominance hierarchies are important under conditions of domestication, for injury during hierarchy formation and poor nutritional condition of low ranking horses are among the most common serious equine behavioral problems . . . ." Although the incidence of serious injuries during hierarchy formation may not be so important in some other farm animals, it is often true that inability to obtain sufficient food because of low status is a real concern for livestock managers.

Priority of Access Information

Dominance relationships may not be at all evident to the human observer until animals are motivated to express those relationships, as when they compete for something that is desired and in short supply. Although priority of access may be extremely helpful in determining status quickly, as in the competitive feeding trials with acquainted horses, it may also fail, as when the hunger drive is so strong as to override normal dominance relationships and cause peck-order violations (King, 1965). These conflicting results indicate that priority of access should be used with caution for indicating status until its association with status under specified test conditions is confirmed, or unless it is the only reasonable explanation available. Some further examples should be useful.

Sexual Differences in Priorities. A complication that needs to be considered is that animals of different sex may have different dominance relationships in different situations. As an example of the heterosexual situation, Tyler (1972) observed that mares which were subordinate to a stallion when hay was provided were likely to bite and kick him when in estrus or when they had young foals with them. Thus, mares were subordinate when competing for feed but behaved as dominants in terms of rejecting unwanted sexual advances and when protecting their young. This type of reversible heterosexual dominance relationship is believed to be rather widespread among species.

Priority of Access to Water (Chickens). Chickens were kept separately in several small flocks, and peck orders were determined by Banks et al. (1979) on the basis of aggressive dominance under conditions of continuous availability of feed and water. The same chickens were then tested for priority of access to feed and water by being placed in a familiar pen where only one of the five within a unisexual group could feed or drink at a time during a 10-minute observation period. All birds were deprived of food and water for about 18 hours before each test period. Athough access to roosting space, dust box, and nest box was also available to only one bird at a time, no relationship of relative dominance and time spent in those places was found. However, the birds were not motivated to use those facilities (males had no need of nests, hens had already laid their eggs, etc.); hence, no relationship would be expected.

Banks and his coworkers found that status was correlated with feeding behavior but not with drinking. Overt competition, including pecks, threats, and avoidances, was seen at or near the feed hopper, and higher-status birds generally had greater feeding activity. (Even so, the association was not perfect; lower-ranking birds in some groups had higher frequencies and duration of feeding.) Surprisingly, no agonistic behavior was seen at the water source. Although thirsty, the birds were orderly at the drinker and status effects could not be detected. The suggestion was made that because birds must first dip their beaks and then raise their heads so that gravity can aid in the passage of water to the gut, eye contact was not possible between individuals, and such contact may be a necessary precedent to pecking and forcing a lower-ranking bird away from water.

Assumed Priority of Access (Swine). The Danes have a government-sponsored swine testing program of long standing and high reputation. For a number of years their standard practice was to test two barrows and two gilts of a particular progeny group by feeding them together and then measuring average daily gain, backfat thickness, and carcass length. It was later decided that feeding the pigs individually would provide better progeny tests. Per Jonsson (1959) analyzed a large amount of data comparing the two methods, with special attention to the relative weight gains of barrows and gilts. The results are of interest because of their implications as to status effects on feeding behavior. Data were not available on individual feed consumption when animals were group fed. It was assumed that differences in daily gain in body weight reflected differences in intake. Results were as shown in Table 12-1.

Of particular interest from the tabular material is the finding that gilts apparently suffered in feed consumption (i.e., they gained less) when they had to compete with the castrated males. When individually fed, the gilts gained faster than the males. (This is an excellent example of a genotype by environment interaction.) Jonsson's report indicates that he believed the barrows were socially dominant and got more than their share when competing with the gilts. This study, and another with chickens (Tindell and Craig, 1959), suggest that more dominant individuals may be stimulated to eat more and may actually perform at higher levels when competing with subordinates than they would otherwise.

What, in fact, can be concluded as to status effects and priority of access to feed when barrows and gilts are fed together? The evidence

Table 12-1   *Effects of keeping and feeding castrated male pigs (barrows) and female pigs (gilts) together in groups of four or separately*

Housing and feeding arrangement[a]	Average daily gain (g) (from 20 to 90 kg body weight)		Barrows minus gilts
	Barrows	**Gilts**	
Groups of four: 2 barrows + 2 gilts	685	670	15 ($P < .01$)
Individually	668	679	–11 ($P < .005$)

[a]Pigs were fed as much as they would eat in 20 minutes.
SOURCE: From results presented by Jonsson (1959).

here is circumstantial. Jonsson believed the barrows were socially dominant, and his casual observations apparently convinced him of that, but on scientific grounds more positive evidence is desirable. It may be that gilts eat as much as barrows in mixed-sex groups but are less efficient in feed conversion in that social environment. Clearly, this kind of experiment is valuable in indicating the actual outcome of the two methods of feeding, but the implications in terms of status effects and priority of access to feed must be tentative until more direct evidence is obtained.

DEVELOPMENT IN YOUNG ANIMALS

Dominance orders cannot develop soon after birth among puppies or kittens because they cannot see, hear, or adequately control their own movements. Precocial species differ in being relatively well developed at birth and would presumably be able to learn dominance relationships within a short time. A general picture of development of dominance orders within the major species follows.

Swine

Several studies (McBride, 1963; Hartsock and Graves, 1976; Scheel et al., 1977) have followed the very rapid establishment of "teat orders" or "nursing orders" among piglets. It is relatively well established within 6 to 8 hours after the completion of farrowing. At first, baby pigs move from nipple to nipple, apparently "sampling" for adequacy of milk supply. Pigs born earlier in the farrowing process tend to be heavier, and having the advantages of earlier arrival and larger size, usually obtain access to the more anterior nipples, which have a greater supply of milk. Agonistic behavior occurs very early and at high frequency as pigs bite, push, and block the approach of others in establishing their claim to a particular nipple. Hartsock and Graves found the frequency of fights per pig to be about eight per hour 2 hours after birth and to drop to about two per hour by 8 hours. Having claimed and defended a particular nipple for several hours, each piglet enjoys a territorial-like advantage thereafter in gaining and maintaining access to it. Late-arriving and "runt" pigs tend to move up and down the udder trying to locate an adequate nipple. They usually nurse from marginally productive posterior teats or fail to gain access (in large litters) and their mortality is high.

When does the dominance order form within litters of pigs? Scheel et al. (1977) believe that it is largely established when baby pigs are competing for productive nipples within the first half-day of birth. Although they saw much agonistic activity shortly after birth, very little aggression was observed among littermates thereafter. They tested for dominance between pairs of littermates after weaning at about 8 weeks of age by observing aggressive behavior while pigs competed for a small amount of applesauce. They found significant correlations between the percentage of fights won by baby pigs at the udder and postweaning dominance within litters. It appears that the relatively peaceful existence among littermates, except when competing for a limited quantity of appetizing food, is a result of their having a well-established dominance order and perhaps also because of socialization during early life.

At weaning time, pigs of different litters are commonly placed together as strangers. Usually, they spend a brief period (15 to 30 minutes) exploring their new surroundings, including strange pigs, by much looking, listening, and sniffing while walking about. The initial exploratory phase is followed by sporadic agonistic activity, which occurs here and there and then seems to spread rapidly, so that nearly all pigs become involved in fighting with strangers that are encountered. Vigorous aggressive activity occurs when 4-week-old weanlings are combined (Figure 12-2) and presumably would also occur at earlier ages.

Figure 12-2 Vigorous fighting occurs during the first day after weanling pigs are put together at 4 weeks of age. (Photograph by the author.)

Fighting gradually decreases among newly assembled pigs, so that little is seen by 24 to 48 hours later, at least in small groups of 6 to 10 (e.g. McBride et al., 1964; Scheel et al., 1977). Nevertheless, when groups of pigs assembled after weaning are fed on a restricted schedule, aggressive incidents may persist at fairly high levels during feeding periods for at least 8 to 10 weeks (Meese and Ewbank, 1972). The latter authors' observations suggested that the mostly linear dominance orders found in groups of eight pigs may not be very stable, as social ranks of some middle- and lower-ranking pigs appeared to change rapidly and without prior warning. However, top-ranking animals maintained their status with only rare exceptions.

If dominance orders are less stable in swine than in most other farm animals, as suggested by Meese and Ewbank (1972), it may be because bidirectional aggression is common in pigs, whereas it is not in most other species. Retaliatory aggressive acts often occur when a subordinate pig is attacked, although they are usually ineffective in reversing dominance. Scheel et al. (1977) reported the unexpected observation that lower-ranking pigs, apparently because of "frustration," would often chase those of higher rank, and as a result, were then usually rebuffed by the chased animal, which reasserted its dominance. McBride et al. (1964) also described bidirectional aggression and agonistic play. Nevertheless, experienced observers are usually confident in their assessment of dominance relationships in pigs.

Dogs and Cats

As with pigs, puppies and kittens are born within litters, but unlike pigs, they are nearly helpless at birth. There is little evidence of puppies "claiming" particular nipples, although kittens do (Ewer, 1961). Apparently little has been established about social organization within litters of kittens, and in later life cats under relatively natural or feral conditions tend to mostly solitary lives within their own territories. Complex dominance relations are expressed when they do come into close contact (summarized by Fox, 1975). Attention will be centered here on the development of dominance relationships within litters of pups and the establishment of such relationships when strange animals meet will be considered briefly.

Development of dominance within litters of pups has been studied extensively by Scott and Fuller (1965), who attempted to determine dominance by laying a meat-covered bone between pairs. All degrees of dominance were observed; occasionally, puppies would share a bone, but more commonly one would seize the bone and appeared to be dominant while possessing it. Although some pups would struggle continually, in other pairs one might seize and hold it for all of the 10-minute test period. An arbitrary decision was made to define dominance by ability to maintain possession for at least 8 minutes. Further information was then obtained by taking the bone from the apparently dominant individual and giving it to the other; repossession of the bone from the apparently subordinate puppy resulted in a classification of "complete dominance." Such tests indicated that dominance had been established in not more than 25% of pairs by 5 weeks in the five breeds tested, but the average for all breeds of complete dominance within pairs was about 50% by 11 weeks. The frequency of complete dominance between littermates was tested again at 15 and 52 weeks with inconsistent results among breeds, but on the average, complete dominance was shown in slightly more than half of all pairs tested. Scott and Fuller pointed out that the dominance tests often caused fighting and may have resulted in dominance being establishd at that time (i.e., they may have hastened the development of dominance relationships).

In considering the within-litter results of competitive food trials with pups as carried out by Scott and Fuller, the question arises as to whether a situation comparable to that seen in chicks and calves (Chapter 11) may not have been at work. Perhaps pups within a litter become moderately tolerant of each other and fail to develop dominance relationships as rapidly as they would if mixed with strange individuals at relatively young ages. However, James (1951), working with two groups in which Terriers and Beagles had been reared together from birth to a year of age, had no trouble in determining dominance in competitive food trials when pairs reared together, but involving the two breeds, were tested; Terriers exhibited their dominant status promptly in controlling the food source.

Breeds of dogs differ greatly in their levels of aggressiveness, presumably because of human selection. Those differences are reflected in the relative complexity of social organization and consequences of dominant and subordinate status in different breeds. Among five breeds compared by Fuller (1953), it was found that individual subordinate animals were likely to be attacked by a "gang" of littermates within the Wire-haired Fox Terrier breed, beginning at about 9 or 10 weeks of age. In some cases the animal attacked was excluded from shelter (in severe weather) and died from exposure; in others it was necessary to remove the victim from the litter to prevent its being killed. Six Terrier puppies were victims among nine litters. Although one Shetland Sheepdog puppy was also attacked in similar fashion (out of five litters), there were no other cases of group attacks within Basenjis (7 litters), Beagles (12 litters), or Cocker Spaniels (12 litters).

Social hierarchies tend to be relatively complex within groups of dogs, but the relationships are generally simpler, more apparent, and more rigidly enforced within highly aggressive breeds. King (1954) compared the relatively aggressive Basenji and tolerant Cocker Spaniel breeds in terms of within-litter social organization and reactions of littermate groups to strangers. (As in most dog studies, small numbers were involved, so that conclusions should be considered as tentative.) The Basenjis, in both male and female groups, formed more rigid hierarchies within litters than Cockers, and males within each breed showed more social dominance than females. Cocker Spaniel males showed very few displays of overt aggressive behavior within their litter and priority of access to desired objectives usually belonged to whichever animal possessed them first. The group of Cocker females studied were so peaceful that no hierarchy could be determined within the litter of three; one was aggressive toward strangers, another was tolerant, and the third member usually stayed away. An observation of considerable interest was that individual strangers of the same breed and sex elicited more aggressive responses than other breed and other sex strangers. The Basenji male group was particularly vicious in attacking strange Basenji males and might have killed them had they not been removed. King proposed that dogs living in organized groups tended to generalize from their own intragroup actions to strangers like themselves; he suggested that animals living in rigid social hierarchies (as in the Basenji groups) might be filled with latent antagonism that would then be directed toward a stranger. On the other hand, the loosely organized Cocker male group was much less aggressive toward strange cocker males, and the Cocker females showed defensive aggression only in repulsing unwanted sexual advances of strange males (no females were in estrus during the study).

Chickens and Turkeys

Newly hatched chicks and poults establish strong primary social relationships from close contact with their mother (if naturally brooded) and with other chicks or other species from prolonged contact during the first few days. By the end of the first week they are very active, easily frightened by strange stimuli and "frolic" or play. By the third week "sparring" or play fighting, usually without physical contact, is common among chicks. Occasional pecking and avoidance of pecking (but without firm dominance relationships being established) become more common. By about the sixth week, some dominance relationships have become established in chicks.

From 6 to 10 weeks peck orders form in small flocks of chicks

(Guhl, 1958). Males become aggressive earlier than females, and by 6 to 8 weeks a peck order is usually present among cockerels; by 8 to 10 weeks it is evident among pullets. In small flocks with both sexes present, a linear or near-linear order usually exists in which the top half is occupied by male and the lower half by female chicks. Relatively few pecks are delivered from cockerels to pullets after the social organization is complete. It appears that the subordinate pullets are not regarded as threatening by the cockerels, and they are relatively tolerant of them. Pecking and threatening frequencies are higher within same sex-groups.

Generally, similar development of peck orders appears to occur among poults, but less detailed observations are available. Young turkeys are believed to develop social hierarchies a few weeks later in age than in chickens.

Sheep and Goats

Among sheep and goats, dominance orders are not very obvious when the animals are grazing adequate pastures and in the non-breeding season. In fact, there is some question as to whether dominance orders are present to any important extent within female groups under such conditions. Crowding into feedlots and competition for concentrated foods produces some aggressive behavior and development of dominance orders; those effects are more obvious in goats than in sheep.

Cattle and Horses

The delayed and reduced aggression and later formation of dominance relationships within groups of heifers reared together as compared to those assembled at 6 and 12 months of age has already been discussed at some length (Chapter 11). Both young foals and calves frequently exhibit aggressive behavior toward other young animals, but stable dominance relationships are not usually established until 6 to 12 months of age.

Tyler (1972) observed among free-ranging ponies that foals only a few days old showed threat postures by laying back their ears and by turning and kicking toward other approaching foals; however, they rarely threatened animals more than a few months older than themselves. Fillies were more aggressive than male foals in frequency of biting and kicking. Dominance relationships tended to become stabilized among young horses during their first winter, but some changes occurred later.

Schloeth (1961) observed that calves of the semiwild Camargue cattle engaged in much playful fighting; sparring and pushing or butting was already frequent at a few days of age, often associated with playful chasing and running away. Of four young heifers that were closely observed from 3 to 4 months old until more than a year later, no clear hierarchy was initially present. Dominance relationships gradually developed among those four, but during the last part of the observation period (when they were 16 to 20 months old), two pair relationships that had been previously established were reversed.

It is particularly interesting to note from Tyler's study that the relative status of young ponies in a natural group is often influenced by their dams'; offspring of high-ranking mothers were especially likely to also attain high rank. Dominant mares were observed on several occasions to come to the aid of their foals, particularly when they were only a few days old. However, older offspring were also aided, although less frequently. As an example, Tyler cites the example of an 8-month-old filly of a very aggressive, dominant mare which threatened ponies as much as 3 years older than herself when competing for hay. Although most of the older fillies returned the threats, three did not and avoided the aggressive daughter of the dominant mare. Those interactions occurred in the absence of the mother; when the filly was near her dominant dam during hay feeds, she increased her status even more. Tyler points out the possibility that inheritance may be involved in such cases, but there may also be opportunity to gain in status through close association with a dominant and aggressive dam and by learning to be aggressive from watching mothers that kick and bite other mares.

FACTORS ASSOCIATED WITH STATUS

Factors influencing the relative status of adult chickens were investigated some years ago at the University of Chicago by Collias (1943). Those results have proven to be generally applicable to other species. Briefly stated, there are two major kinds of influences: (1) developmental and physiological, and (2) psychological. Among developmental and physiological factors associated with high rank, Collias found that males are generally dominant to females, genetic stocks differ in dominance ability, and older, larger, and stronger individuals (the last three tend to be highly correlated) have greater success in attaining dominance.

Psychological influences can be important in determining the outcome of agonistic encounters between strangers. Individuals have greater success in aggressive encounters when they are in familiar surroundings ("home-court" advantage) and among acquaintances, except for the stranger. Previous experience in winning and losing can also be of major importance; coaches of prize fighters and of athletes engaging in competitive sports are well aware of such effects. The psychology of success or losing in animals may, however, be of relatively short duration.

Size and Genetic Influences on Status in Cows: An Example

Observations of Wagnon et al. (1966) indicated some factors influencing status of beef cows in a California herd. Their animals were selected from a larger group formed from numerous sire families and bloodlines. All cows were without horns. (The presence or absence of horns are of major importance in determining status when strange cows meet; Bouissou, 1972.) The cows were apparently acquainted prior to being drawn from the larger herd and had weaned their second crop of calves. Their average age was 46 months and 10 were sampled from each of the Angus, Shorthorn, and Hereford breeds. Those 30 animals were placed in a 2-ha (5-acre) lot and fed at a 66-foot-long fence-line manger (allowing 2.2 feet or about 67 cm of feeder space per cow). The feeder was shorter than required so that not all animals could feed at the same time. Therefore, they competed for feeder space. Feed was provided on the basis that none was left overnight. The group was kept together over a 2-year period and cows were ranked for social dominance on the basis of percentage of other cows dominated. Average weights and social ranks by breeds are given in Table 12-2. Herefords were the heaviest breed but were

Table 12-2   *Average body weights and social ranks of Angus, Shorthorn, and Hereford cows when kept together over a 2-year period*

Breed group[a]	Body weight (lb)		Social rank[b]	
	1964	1965	1964	1965
Angus	1058	1214	8.7	9.0
Shorthorn	994	1203	17.4	16.4
Hereford	1083	1248	20.4	21.0

[a]Each breed represented by a sample of 10 cows.
[b]Social ranks of individual cows ranged from 1 (most dominant) to 30 (most subordinate).
SOURCE: From results presented by Wagnon et al. (1966).

lowest in dominance status both years; breed differences were highly significant both years. Here is a clear example of genetic influence at work. This influence was obviously of greater importance than body size. Nevertheless, body size was important in determining rank *within* breeds, where it was determined that a significant association existed (heavier cows ranked higher).

Another result of interest was that the social ranks of the three breeds and also particular cow's ranks were highly repeatable from one year to the next. This result indicates a high level of social inertia; once a group is organized it tends to stay the same.

Social Inertia

Dominance orders tend to be relatively stable, especially when females are involved. Social inertia occurs when individuals learn their relationships thoroughly, so that they promptly show dominant or submissive behavior when they meet. Ordinarily, it is assumed that a group is highly organized after fighting becomes uncommon and physical contacts are replaced by threats and passive submission. Outward signals of status may then become very subtle and difficult for the human observer to detect except when a scarce resource is involved. Large differences in age, size, strength, genetic background, and previous experience are powerful stabilizing influences on social organization.

Group Size. It has already been indicated that larger groups of hens in cages have higher frequencies of agonistic acts during the first several weeks after they are placed together, even when area per hen is the same for larger and smaller groups (Al-Rawi and Craig, 1975; Al-Rawi et al., 1976). It is reasonable to postulate that part of the higher frequencies in larger groups is caused by longer periods being required for thorough learning of dominant or subordinate roles relative to every other individual in the larger groups. Peck-order violations certainly occur with greater frequency in larger groups as demonstrated by Banks and Allee (1957). Nevertheless, Al-Rawi et al. (1976) found that the frequency of aggression eventually declined in their larger groups, perhaps partly because all hens became fully acquainted with their cagemates and errors of recognition of individuals and their status became negligible, so that reinforcement of rank by aggressive acts of dominants was no longer required. Third-party effects were probably also important in the high-density environment used (Ylander and Craig, 1980).

What about stability of individual status and social inertia in groups much larger than those found in the evolutionary environment of a species? There is little direct evidence on this because of the very large number of pair relations possible in sizable groups which makes determination for all possible pairings nearly impossible. Thus with 10, 100, and 1000 in a group, there are 45, 4950, and 499,500 pair relationships possible, respectively. As already indicated (Chapter 11), when groups of cattle are combined, family units or animals acquainted from early in life attempt to maintain contact. Also, when strangers are placed together in relatively large groups and areas, they tend to live in localized areas, at least in flocks of hens. Those observations suggest that large groups may, in reality, consist of a collection of subgroups. Although the evidence is indirect, the finding that frequency and severity of agonistic acts of hens did not differ among flocks of 100, 200, and 400 (Craig and Guhl, 1969) and that it was quite comparable to that observed in flocks of 25 hens (Craig et al., 1965) suggests that the larger groups were actually socially organized on a "neighborhood" basis. Other evidence in the Craig and Guhl study also supports this view.

Temporary Absence. How long will an individual be remembered if it is absent from the group? Chickens forget each other if kept apart for 2 or 3 weeks, as indicated by subordinates attacking those that were formerly dominant to them in the peck order.

Duration of memory for individual pigs may depend on their relative status in the dominance order. When individuals were removed for varying intervals and then returned to the group, it was learned that top-ranking pigs could be kept away for more than 25 days without being attacked, but low-ranking pigs were severely attacked after an absence of only 3 days (Ewbank and Meese, 1971).

It is not clear how long animals of other species remember each other. In the classical literature we read that Ulysses' dog Argus remembered his master after a 10-year absence. Cows moved from group to group are often involved in agonistic activity when they join a new group, but the duration of memory of cattle for other individuals is not clear at this time.

Difficulty of Recognition. The role of vision in recognition among hens must be large as shown by Guhl and Ortman (1953). Interestingly, they found that altering the appearance of the head and neck was far more important than changes over the rest of the body. Changing the comb from lopping from left to right (or vice versa) was enough to cause subordinates to attack a dominant penmate returned after a short absence. Spectacular changes over the rest of the body, such as gluing red feathers on a white hen or adding male "sickle" feathers to the tail of a hen, had little effect (see Figure 5-2).

The importance of vision and of physical contact in maintaining dominance relationships and control of a food source between pairs of 3-year-old heifers has been examined by Bouissou (1971). She used a group of eight that had been together since 6 months of age. Optical information was deprived by placing linen masks over the entire head, except for the horns, ears, and muzzle. When pairs of hungry heifers had a single bucket of feed placed before them for a 3-minute test, the dominant animal always controlled the food source for the entire period; however, when masked, the subordinates were usually able to feed (in 23 of the 28 pairings), but on the average gained access for only one-fourth of the test period. It was observed that when heifers were masked, aggressive acts were always in the same direction and similar to those seen when the animals were tested in the control situation. Only the dominant member was aggressive, and the interaction consisted of either a light or sometimes a more violent thrust with one horn, with the result that the subordinate was evicted, as shown in Figure 12-3. Because the animals could not see each other and were not communicating by sound, Bouissou concluded that the sense of smell was the means of recognition involved. Nevertheless, recognition by odor alone was obviously not adequate for maintaining control of the food source in most pairings, as subordinates were usually able to gain temporary access to the food until forced away by an aggressive act. Without masks, aggressive hornings by the dominant were usually not required in maintaining control; visual threats were adequate.

When the same set of eight heifers were separated in a familiar pen by a partition consisting of vertical bars, but having an opening in the center where a bucket of feed could be placed, the results were quite different. In this situation they could see and smell each other, but physical contact was limited to their muzzles only. When hungry pairs were fed in this situation, dominant and subordinate members averaged the same amount of time feeding. Although dominant members tried to evict subordinates by pushing with their muzzles, they were unsuccessful. Threatening postures by the dominants were also ineffective. Bouissou (1971) pointed out that although physical contact is not usually required when a dominant feeds from a limited source in the presence of a subordinate heifer, it must be perceived as possible for feeding inhibition to occur. Although status effects were temporarily suspended between heifers when physical contact was perceived as impossible, Bouissou notes

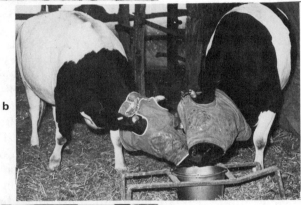

Figure 12-3 When hungry heifers competed for feed, dominant individuals always controlled the food source for the entire test period under "control" conditions, but subordinate heifers sometimes gained temporary access when both were deprived of optical information. (a) Dominant heifer prevents subordinate from feeding. (b) Dominant heifer supplants subordinate during feeding competition. (c) Subordinate animal stands aside while dominant feeds. (From Bouissou, 1971.)

that her results differ from those obtained in apes in an analogous situation. It will also be shown later that when a dominant ram is nearby, subordinate rams are inhibited in breeding activity, even if physical contact is impossible. Clearly, those conditions under which status effects are maintained are modified according to how the animals recognize each other or perceive that aggressive acts can or cannot be delivered.

M. F. Bouissou related, in a professional meeting, how a cow that had fallen in a mudhole and been rather thoroughly covered was suddenly treated as a stranger by her acquaintances. Although her appearance was suddenly changed, it may be that her normal body odors were also no longer evident, thereby making recognition doubly difficult.

Sudden changes in posture or unusual movements may cause temporary loss of recognition. Status can have significant effects on frequency of mating in most farm animals (Chapter 17). This seems to be brought about in part because of interference by dominant males with subordinate males during the mating act. Subordinates rarely interfere with a dominant male during mating in mammals. In view of that situation, the observations of Kratzer and Craig (1980) with cockerels are somewhat surprising; they found that matings were interfered with (at a certain frequency) regardless of the status of the mating male. As indicated above, chickens (and probably other poultry) rely largely on visual recognition. Therefore, it was suggested that sudden changes of posture of a cockerel while mating may alter the ability of other males to recognize him.

Changes in Dominance Potential. Fully mature females within undisturbed, socially organized groups tend to remain in the same rank over periods of months or even years, at least within herds of horses and cows and flocks of hens. It seems likely that much the same situation applies in other domesticated species.

Guhl (1953) investigated the repeatability of peck-order status. Pullets that developed peck orders as they matured were isolated at a later age and then reassembled after several weeks. Because chickens forget each other rather rapidly, they behaved as strangers when reassembled. Among four such flocks tested there was a significant correlation between early and later peck orders in only one. Some correspondence existed for rank in the two peck orders in the other flocks, but changes in dominance relationships were found for a number of pairs. It appears likely that pullets which mature earlier tend to be higher in peck orders established as birds develop (Bhagwat and Craig, 1977).

Further evidence has been collected on the stability of social status in flocks of hens in which changes in social dominance potential occur. Six diverse strains of chickens were assembled into intermingled-strain and separated-strain flocks by Tindell and Craig (1959). Hens in three intermingled flocks (48 hens in each) were observed and relative social status was determined (between 5 and 9 months of age) for the strains. Significant differences were found among the strains and strain status continued at the same level over the 4 months. At 8 months of age, hens were removed temporarily from separated-strain flocks and pairs of hens from different genetic stocks were matched in "initial pair contests" by Bellah (1957). Surprisingly, the strain that was lowest in status in flocks assembled at 5 months had become highest in ability to win pair contests with strangers by 8 months of age. Nevertheless, that strain continued in lowest status in the intermingled flocks. Social inertia had effectively locked it into that position.

Observations of semiwild cattle by Schloeth (1961) indicated that essentially three separate dominance orders were present within a herd made up of both sexes and all age groups. In a sense there was only one large dominance order with adult bulls in top positions, adult cows in the middle, and juveniles at the bottom. Adolescent bulls did not fit neatly into the general pattern, however. They began to work their way up through the adult cow hierarchy at about $1\frac{1}{2}$ years of age and emerged at or near the top of the mature cow group at about $2\frac{1}{2}$ years. At that time they occupied the lowest positions of the adult male social hierarchy. Adult bulls seldom behaved aggressively toward cows, but adolescent bulls were in relatively frequent contests with cows as they moved up in the hierarchy. Clearly, this kind of social situation must be common among group-living species in which polygyny occurs, and there is marked secondary sexual dimorphism, with males being much larger and more powerful after sexual maturity.

REFERENCES

AL-RAWI, B. A., and J. V. CRAIG, 1975. Agonistic behavior of caged chickens related to group size and area per bird. Appl. Anim. Ethol. 2:69–80.

——, 1976. Frequency of decisions in initial pair contests: effects of sex, observer, and feeding stimulus. Poult. Sci. 55:462–464.

——, and A. W. ADAMS, 1976. Agonistic behavior and egg production of caged layers: genetic strain and group-size effects. Poult. Sci. 55:796–807.

BANKS, E. M., and W. C. ALLEE, 1957. Some relations between flock size and agonistic behavior in domestic hens. Physiol. Zool. 30:255-268.

BANKS, E. M., D. G. M. WOOD-GUSH, B. O. HUGHES, and N. J. MANKOVICH, 1979. Social rank and priority of access to resources in domestic fowl. Behav. Processes 4:197-209.

BELLAH, R. G., 1957. Breed differences among chickens as related to compatability when reared together. M.S. thesis, Kansas State University Library.

BHAGWAT, A. L., and J. V. CRAIG, 1977. Selecting for age at first egg: effects on social dominance. Poult. Sci. 56:361-363.

BOUISSOU, M. F., 1971. Effet de l'absence d'informations optiques et de contact physique sur la manifestation des relations hiérarchiques chez les bovins domestiques. Ann. Biol. Anim. Biochem. Biophys. 11:191-198.

——, 1972. Influence of body weight and presence of horns on social rank in domestic cattle. Anim. Behav. 20:474-477.

COLLIAS, N. E., 1943. Statistical analysis of factors which make for success in initial encounters between hens. Am. Nat. 77:529-538.

CRAIG, J. V., and A. M. GUHL, 1969. Territorial behavior and social interactions of pullets kept in large flocks. Poult. Sci. 48:1622-1628.

CRAIG, J. V., L. L. ORTMAN, and A. M. GUHL, 1965. Genetic selection for social dominance ability in chickens. Anim. Behav. 13:114-131.

EWBANK, R., and G. B. MEESE, 1971. Aggressive behaviour in groups of domesticated pigs on removal and return of individuals. Anim. Prod. 13:685-693.

EWER, R. F., 1961. Further observations on sucking behavior in kittens, together with some general considerations of the interrelations of the innate and acquired responses. Behaviour 17:247-260.

FOX, M. W., 1975. The behaviour of cats. *In* E. S. E. Hafez (Ed.), The Behaviour of Domestic Animals, 3rd ed. The Williams & Wilkins Company, Baltimore.

FULLER, J. L., 1953. Cross-sectional and longitudinal studies of adjustive behavior in dogs. Ann. N.Y. Acad. Sci. 56:214-224.

GUHL, A. M., 1953. Social behavior of domestic fowl. Kans. Agric. Exp. Sta. Bull. 73.

——, 1958. The development of social organisation in the domestic chick. Anim. Behav. 6:92-111.

——, and L. L. ORTMAN, 1953. Visual patterns in the recognition of individuals among chickens. Condor 55:287-298.

HARTSOCK, T. G., and H. B. GRAVES, 1976. Neonatal behavior and nutrition-related mortality in domestic swine. J. Anim. Sci. 42:235-241.

HOUPT, K. A., K. LAW, and V. MARTINISI, 1978. Dominance hierarchies in domestic horses. Appl. Anim. Ethol. 4:273-283.

JAMES, W. T., 1951. Social organization among dogs of different temperaments, Terriers and Beagles, reared together. J. Comp. Physiol. Psychol. 44:71-77.

JONSSON, P., 1959. Investigations on group versus individual feeding and on the interaction between genotype and environment in pigs. Acta Agric. Scand. 9:204-228.

KING, J. A., 1954. Closed social groups among domestic dogs. Proc. Am. Philos. Soc. 98:327-336.

KING, M. G., 1965. Disruptions in the pecking order of cockerels concomitant with degrees of accessibility to feed. Anim. Behav. 13:504-506.

KRATZER, D. D., and J. V. CRAIG, 1980. Mating behavior of cockerels: effects of social status, group size and group density. Appl. Anim. Ethol. 6: 49-62.

LOWRY, D. C., and H. ABPLANALP, 1970. Genetic adaptation of White Leghorn hens to life in single cages. Br. Poult. Sci. 11:117-131.

——, 1972. Social dominance difference, given limited access to common food, between hens selected and unselected for increased egg production. Br. Poult. Sci. 13:365-376.

MASURE, R. H., and W. C. ALLEE, 1934. The social order in flocks of the common chicken and the pigeon. Auk 51:306-327.

McBRIDE, G., 1963. The "teat order" and communication in young pigs. Anim. Behav. 11:53-56.

——, J. W. JAMES, and N. HODGENS, 1964. Social behaviour of domestic animals. IV. Growing pigs. Anim. Prod. 6:129-139.

MEESE, G. B., and R. EWBANK, 1972. A note on the instability of the dominance hierarchy and variations in level of aggression within groups of fattening pigs. Anim. Prod. 14:359-362.

SCHEEL, D. E., H. B. GRAVES, and G. W. SHERRITT, 1977. Nursing order, social dominance and growth in swine. J. Anim. Sci. 45:219-229.

SCHJELDERUP-EBBE, T., 1922a. Beiträge zur Sozialpsychologie des Haushuhns. Z. Psychol. 88:225-252.

——, 1922b. Contributions to the social psychology of the domestic chicken. (English translation by M. Schleidt and W. Schleidt from Z. Psychol. 88: 225-252.) In M. W. Schein (Ed.), Social Hierarchy (Benchmark Papers in Animal Behavior, Vol. 3). Halsted Press, A Division of John Wiley & Sons, Inc., New York.

SCHLOETH, R., 1961. Das Sozialleben des Camargue-rindes. Qualitative und quantitative Untersuchungen über die sozialen Beziehungen — insbesondere die soziale Rangordnung — des halbwilden französischen Kampfrindes. Z. Tierpsychol. 18:574–627.

SCOTT, J. P., and J. L. FULLER, 1965. Genetics and the Social Behavior of the Dog. University of Chicago Press, Chicago.

SERENI, J. L., and M. F. BOUISSOU, 1978. Mise en évidence des relations de dominance-subordination chez le cheval, par la méthode de compétition alimentaire par paire. Biol. Behav. 3:87–93.

TINDELL, D., and J. V. CRAIG, 1959. Effects of social competition on laying house performance in the chicken. Poult. Sci. 38:95–105.

TYLER, S. J., 1972. The behaviour and social organization of the New Forest ponies. Anim. Behav. Monogr. 5:85–196.

WAGNON, K. A., R. G. LOY, W. C. ROLLINS, and F. D. CARROLL, 1966. Social dominance in a herd of Angus, Hereford, and Shorthorn cows. Anim. Behav. 14:474–479.

YLANDER, D. M., and J. V. CRAIG, 1980. Inhibition of agonistic acts between domestic hens by a dominant third party. Appl. Anim. Ethol. 6:63–69.

13

Feeding Problems and Vices

UNFAMILIAR FOOD AND DEVICES

Given a choice, animals usually prefer to continue eating those foods with which they are already familiar. This is adaptive because a very limited sampling of novel foods allows the animal to learn, by physiological responses and associated feelings of well-being or discomfort, whether the food is good for it or not. Nevertheless, nutritious foods may not be eaten because of poor palatability or unfamiliarity to the point that normal growth or functioning may not be attained if they are the only ones available or abundant.

Early exposure to types of food that may need to be used later in life can result in its ready consumption when needed. Ensminger (1976, in Kilgour, 1978) indicated that Chinese farmers feed a water hyacinth extract, together with sow's milk, to pigs shortly before weaning. Pigs exposed in this way learn to like water hyacinth and eat them readily later in life. This example indicates how a plant that is ordinarily considered to be a nuisance may be harvested for a useful purpose.

Cattle and sheep are known to develop particular preferences for pasture plants and may have difficulty in changing over from diets consisting entirely of roughages to concentrated feeds or in accepting supplementary concentrates while on pasture. In an Australian study, Arnold and Maller (1977) brought sheep to the same pasture from different geographical locations. Adult sheep took longer than lambs to adjust. There were large differences in grazing preferences among groups coming from different areas, but in general sheep from all areas adapted adequately. Arnold and Maller (1974) found that feeding some grain early in life overcomes the reluctance of sheep to eat it later. Some animals simply do not eat supplementary feed given on pasture. Part of the problem could be caused by social competition.

By using sophisticated labeling techniques it is now possible to determine which animals on pasture have little or no intake of supplementary feeds. Thus, it was determined that about half of 200 sheep were rejecting liquid containing urea in molasses (Nolan et al., 1975). Results vary greatly among groups of cattle in the proportion that totally reject liquid supplements or supplements incorporated into blocks to be licked. From 8 to 82% of all animals totally rejected supplements in nine comparisons reviewed by Lynch (1980) from studies by J. V. Nolan, R. A. Leng, and their coworkers. Lynch suggested that feeding competition, low palatability, and lack of experience with supplementary feeds are all possibly important variables.

Animals may have difficulty in accepting the same food when presented in a different kind of feeding device or physical form. Several studies have shown that lambs and calves accustomed to drinking their liquid diets from artificial nipples or buckets exclusively had difficulty in locating their food or feeding by the alternative method; some required extensive help by caretakers or became dehydrated and needed replacement therapy (reviewed by Lynch, 1980). Those working with chickens know that birds not accustomed to a particular kind of watering device or to reaching water by extending heads and necks through openings in wire pen fronts can become dehydrated and die of thirst. Young pigs not accustomed to feeders that require lifting of a cover occasionally fail to find it until assisted. Careful observation can indicate whether a problem in finding or accepting food or water is present, so that remedial measures can be taken if needed. As soon as one animal of a group has learned to use a feeding or watering device, others are likely to be attracted and learn to operate it also. Adding new animals to a facility that already has one or more experienced individuals present often prevents problems with unfamiliar feeding or watering devices.

INDIVIDUAL FEEDING
AND SOCIAL FACILITATION

When animals feed in groups, two kinds of social influence are operative: social facilitation (one form of allelomimetic behavior) and agonistic behavior. Social facilitation increases feeding, whereas agonistic behavior is likely to reduce intake by subordinates.

Individual feeding of animals may be useful in several ways. It may be physiologically and economically important for specific requirements associated with age, stage of development, level of production, pregnancy, or activity. Breeders of species used for food production have a special interest in increasing efficiency of feed use by their stock; individual feeding is one means of measuring that efficiency so that selection may then take place. Nutritionists may wish to use individual feeding of animals in special crates or chambers so that relatively exact measures of metabolic changes can be obtained.

When two or more feed together, the quantity eaten may be increased by increased stimulation to eat or by reducing anxiety if animals are in an unfamiliar environment or fear-producing situation. The importance of social facilitation in increasing feed intake varies among species; in some, the other individual must be physically present for maximum consumption, in others, it may be sufficient that the other can be seen, heard, or perhaps even smelled.

Cattle and Sheep

Hereford steers and heifers were wintered on early- or late-cut hay by Kidwell et al. (1954). In each of 3 years the cattle under test were subdivided so that some were penned together while others were placed in comparable pens by themselves. Animals separately penned were kept apart from those in adjacent pens only by wire fences. Nevertheless, individually penned steers and heifers gained significantly less when fed the better-quality, early-cut hay. Animals fed the lower-quality, late-cut hay did not differ in gains. The results are shown in Table 13-1.

Although feed intake was not reported in Kidwell's study, it seems safe to assume that most, or all, of the significant differences

Table 13-1 *Effects of group and individual feeding on gains of steers and heifers*

| | Average daily gain[a] (lb) | | | |
| | Early-cut hay | | Late-cut hay | |
Sex	Group fed	1/pen	Group fed	1/pen
Steers	1.01**	0.67	0.34[(ns)]	0.34
Heifers	0.37*	0.00	-0.24[(ns)]	-0.23

[a]Probability that differences between group fed and individually fed heifers or steers receiving the same ration differed by chance alone is indicated by: ns = not significant (could be due to chance); * = $P < .05$; ** = $P < .01$.

SOURCE: After Kidwell et al. (1954).

in daily gain were caused by differences in feed consumed. Kidwell observed that the individually penned animals were restless, nervous, and apt to waste their hay. Either anxiety caused by physical isolation or the absence of stimulus provided by a feeding companion appears responsible. When the lower-quality hay was fed, the differences in daily gain disappeared. In this case the animals were probably consuming their hay at a "survival" level.

Dairy calves are commonly separated from their dams soon after birth (usually less than 3 days), with satisfactory results if colostrum is fed initially. Calves may be kept in small, individual pens and encouraged to consume high-quality hay and concentrates so that their liquid diet may be discontinued at an early age. Warnick et al. (1977) compared the weight gains and number of days required before calves were eating concentrates when kept in groups of six, in individual hutches, and when isolated visually as well as physically from other calves. Although weight gains to 10 weeks of age were somewhat less for the isolated calves, the difference was not significant. However, social facilitation was important in learning to consume dry feeds. Those kept in groups were eating the desired amount of concentrate (0.45 kg daily) by 22 days, calves in individual hutches required 29 days, and isolated calves did not consume that much until 37 days. Warnick and his colleagues suggested that "exploratory and imitative behavior likely was enhanced in the group environment and resulted in calves learning to feed earlier. . . . "

Experience with cattle kept in metabolism crates indicates that physically isolated individuals usually consume only about 50 to 60% as much per animal as when two or more are kept together; sheep are even more susceptible to isolation (L. Harbers, personal communication). Those results are obtained even when other animals are present in the same room. As in Kidwell's study, the difference in feeding activity between individually and group-fed cattle in metabolism crates decreases when low-quality feeds are given.

Chickens

It was shown more than half a century ago (Katz and Revesz, 1921) that an apparently satiated hen would feed again when a hungry bird was introduced. Commercial egg-strain hens, kept in individual cages but having audiovisual contact with others, eat enough to sustain high levels of egg production and sometimes need to have their intake restricted to avoid laying down excessive body fat.

When Hughes (1971) continuously observed adolescent chickens in cages during 9-hour days (food was removed during the dark hours), he found synchrony of feeding; either fewer or more were generally feeding at any one time than would be found if the time of feeding was on a random basis.

Dogs

It is well known that some pet dogs lose weight when the family is away even though a good-hearted neighbor puts the usual food in the animal's feeding bowl. Anxiety and reduced feeding associated with separation from the family "pack" is the most likely explanation.

Swine

Social facilitation of feeding behavior is commonly observed in pigs; when one walks to the feeder and begins feeding, others usually approach and begin to feed also. The sound of feeding often appears to be an adequate stimulus. Physical presence of another within the same pen (or metabolic crate) is not required as long as other pigs are nearby. Danish swine testing station results indicated that gains in weight are essentially equal for individually fed and group-fed pigs.

Kilgour (1978) indicated the importance of considering social facilitation of feeding among nursing pigs in the design and management of a rearing house. A newly built house was to be used continuously, with sows and their litters being brought in at 3 days of age and removed at 28 days when pigs were weaned. It was soon discovered that the weaned pigs lost weight for several days while adjusting to solid feed. The same feed had been available to the piglets while they remained with the sow, but was not consumed. Sows were found to be in full lactation at weaning. What had happened? Observations indicated that all pigs in the house were suckling at the same high frequency as were the 3-day-old pigs. When the youngest pigs squealed to be fed, it was taken up by older pigs as well, and all sows rolled over and allowed their young to suckle. Kilgour wrote that a farmer, hearing of this finding, kept his weaned pigs next to the nursery barn and found that they would go to the feeder and begin eating whenever they heard pigs inside the nursery making sounds associated with nursing. This intriguing report suggests variations in pig management that might be useful.

STATUS AND FEEDING

Competitive feeding trials have been used widely for determining dominance relationships in domesticated animals (Chapter 12). In those situations animals are usually deprived of food for longer periods and then fed for shorter intervals than would be encountered in normal husbandry. What about everyday situations in which animals are fed on a regular, but somewhat limited basis and also when feed is available continuously?

Beef Cows

Wagnon (1965) observed a cow herd on a California range during supplemental feeding. His animals were of all ages, from 2- to 10-year-olds. It was obvious that the youngest cows were having trouble getting their share. Average weight changes associated with the supplemental feeding period are shown for 2-year-old heifers during two 4-year periods. During the first 4 years heifers were kept with older animals; during the second they were pastured and fed separately.

Weight changes of 2-year-old heifers

Treatment	Change in weight
Pastured and fed with older cows	Lost 25 lb (11.4 kg)
Pastured and fed separately	Gained 46 lb (20.9 kg)

After removal of the 2-year-old heifers it was noted that 3-year-olds were also less competitive in feeding than older cows. Subsequent separation and combining of 2- and 3-year-olds worked well. Although a dominance order was present, it was evident that there was less aggression than found among older cows.

In their study of beef-cow behavior in a dry-lot situation, Wagnon et al. (1966) included observations of cows that were unable to feed because of inadequate space at the feed bunk (they waited their turn) or because they were "butted out" of a feeding position. Breed social status was associated with inability to gain or maintain access to feed, as shown in Table 13–2. The low-status Hereford cows were low in feeding rights; although they composed only one-third of the herd, they were represented by 45% of all cows observed waiting to feed and were forcefully displaced from the feeder far more frequently than were cows of the more dominant Angus and Shorthorn breeds.

A few dominant cows were observed to move alongside the feed bunk, forcing others away until they either reached a feeding site which they favored or until they were stopped by the presence of a higher-ranking cow. Most cows attempted to gain access to the feed bunk by moving in between others.

Table 13-2 *Breed social-status effect on feeding behavior in a competitive situation*

| Breed | Breed status | Cows unable to feed (total observations) | |
		Waiting to feed	Butted out
Angus	Top	478	9
Shorthorn	Middle	406	11
Hereford	Bottom	714	100

SOURCE: After Wagnon et al. (1966).

Serious feeding problems were found among mature beef cows in a study by Schake and Riggs (1972) and indicate how new developments in management can lead to unexpected stress and related poor performance. This study was set up to examine a confinement cow–calf production system. Mature Hereford cows, 7 to 13 years old, were obtained from four Texas ranches; all were in midgestation and had been dehorned as calves. Initially, they were placed in four groups (by source) for a 2-week period and accustomed to the ration to be fed. They were then divided into six new herds of 11 cows each, with approximately equal numbers of cows from each source being assigned to each of the new groups. A high level of persistent agonistic activity occurred over the next 4 days as new social hierarchies were established. Over 10% of the cows became "totally submissive" and would not compete for feed. Schake and Riggs became so concerned for the welfare of the submissive cows that they were fed separately, but were then returned to their herds for the remainder of the day. After 2 weeks, half of the submissive cows would feed with their herdmates. The others required 2 to 3 more weeks of individual feeding before they would compete for feed within the group.

Further results of the Texas study indicated that although low-ranking cows were finally able to feed with the group, their feed intake was clearly reduced, as shown by the example in Figure 13-1, and their performance suffered. Prior to calving, four cows of each herd were removed for other purposes. The seven remaining cows were categorized for comparative purposes as top-, middle-, and bottom-status groups with two, three, and two individuals per classification within each herd. Although three feeding levels were used (two herds per level), I have combined the results for the six herds and compared the relative performance of the middle and bottom groups with those of the top-status group (taken as 100% performance) in Table 13-3.

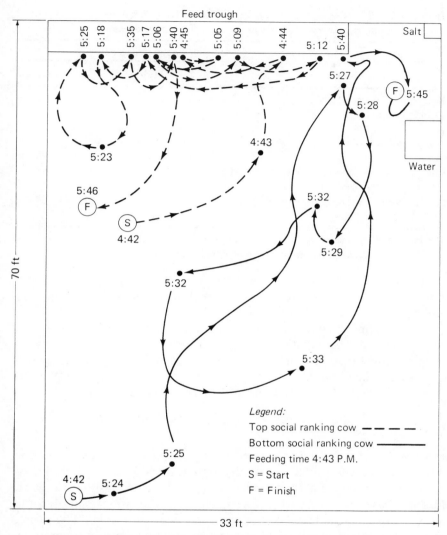

Figure 13-1 One-hour feeding pattern for two cows of different social rankings. (From Schake and Riggs, 1972.)

Beef Steers

Two-and 3-year-old Hereford and Brahman X Hereford crossbred steers kept in groups of six and fed low-quality hay (drought feeding conditions) were observed by McPhee et al. (1964). They found fairly stable linear social orders. Although high-ranking animals had

Table 13-3   *Effect of social status on productivity of range beef cows after strangers were assembled in a confined environment*

| | Social-status category | | |
Productivity item	Top	Middle[a]	Bottom[a]
Number of cows/category within each group	2	3	2
Total cows	12	18	12
Cow weight, last month of lactation	100	90	77
Milk yield, 24 hr, last month	100	100	68
Calf weight at 205 days	100	95	75
Cows pregnant at weaning of calves	100	79	54

[a]Productivity items are expressed as percentage of top social-status group.
SOURCE: Calculated from results presented by Schake and Riggs (1972).

priority at the feed trough, no relationship was found between rank and growth.

There appear to be no studies on the relationship of status and feeding in the large cattle feeding operations that are of major economic importance in North America. However, it is apparent that at least one behavioral problem is present in those feedlots (the "bullersteer syndrome"), which will be considered in Chapter 17.

Dairy Cows

Dairy cattle may be fed by presenting feed at several times during the day with complete consumption of all that is given each time, or alternately by having feed available on a 24-hour basis. Bosc et al. (1968) reasoned that with fodder available at all times, subordinate heifers might be able to obtain a more adequate intake. Therefore, they set up a study with 14 two-year-old heifers, kept in open housing, and fed in the two ways. The animals were observed continuously during the day (8 A.M. to 7 P.M.) for agonistic behavior, and the amount of time each spent feeding was recorded. Two of the 14 were socially disadvantaged; of those two, one received nearly one-third and the other nearly one-half of all aggressive acts, while the remaining animals received only 0.2 to 3.9%. When feed was given several times daily, animals of the dominant group averaged nearly 4 hours at the rack and the two subordinates slightly less than 1½ hours. When feed was continuously available, the dominants averaged about

4½ hours and the subordinates slightly more than 1½ hours. (Additional time was spent eating roughage and litter.) Therefore, when feed was continuously available, the subordinate pair increased their time at the feed rack slightly, i.e., by 20 minutes daily, but an important difference persisted, with the dominants spending even more time there. Although the rack was large enough for about 12 animals to feed at any one time, they obviously did not do so, but avoided close contact while feeding.

Bouissou has explored the effect of social dominance on feeding of dairy cows when physical contact is possible, when animals are blindfolded, or when separated partially or completely by barriers through which they can see, hear, and smell (Bouissou, 1970, 1971). As discussed earlier (Chapter 12), loss of optical information allowed subordinates to feed briefly, but dominance relations were maintained; however, a physical barrier allowing muzzle contact only prevented the expression of feeding rights at a feed source. Various kinds of barriers for use with side-by-side feeding have also been tried, with results as shown in Figure 13-2. Bouissou suggests that subordinate heifers have a "notion of protection" when such barriers are present, particularly when head contact is prevented. Thus, with no barrier present, hungry subordinates fed only 7 seconds during 3-minute test periods, while head protection allowed subordinates to feed for more than 2 minutes from a 70-cm (28 inch)-long trough. Even a single horizontal bar above the trough provided a considerable notion of protection, as subordinates fed for nearly half of the test period when it was present.

Dairy cows in heavy milk production may be allowed free and continuous access to a complete, balanced ration. Friend et al. (1977) reported that under such conditions, in a small experimental herd, average feeding time was not reduced until less than 20 cm (8 inches) of feeder space was allowed per cow. Social dominance was significantly associated with time at the feed bunk when 20 cm space or less was available for each cow. Friend and his coworkers estimated, from chemical techniques, that actual time spent with the head over the feed bunk was a poor indicator of the amount of feed consumed. Cows spending intermediate amounts of time at the feeder apparently ate the most.

The importance of social status in determining feeding behavior and nutrient intake of dairy cows appears to be somewhat controversial at the present, partly perhaps because feeding time itself is a relatively unreliable indicator of actual feed consumption. With continuous availability of feed, it would seem likely that dairy cows of low status would have ample opportunity to feed, yet Bosc et al.

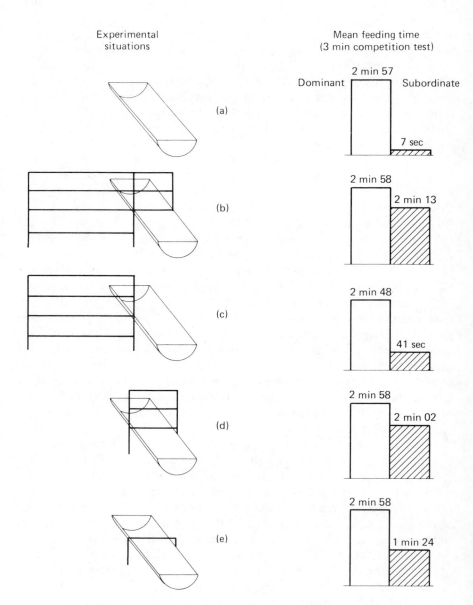

Figure 13-2 Effect of different types of physical barriers on side-by-side feeding time of hungry cows classified according to dominance status. Barriers providing complete protection of the head (b and d) allowed subordinates to feed for more than two-thirds of the test period, whereas partial protection (e) allowed feeding for about half of the period. (Prepared by M. F. Bouissou from results presented in Bouissou, 1970.)

(1968) found that very low status heifers spent much less time at the feed rack than dominant ones, even with rather generous space allowance. On the other hand, the results of Friend et al. (1977) indicated that status would be of only limited importance unless feeding space was severely restricted. Differences in feeding equipment and feedstuffs may be responsible for such discrepancies. Feeding fodder (hay) in a rack is considerably different from feeding complete feeds (silage and grain mixed together) and placed in a feed bunk.

In a later report, Friend and Polan (1978), using multiple-regression analysis, concluded that dominance values were not adequate indicators of priority at the feed trough. This conclusion was challenged by Beilharz (1979), who pointed out from Friend and Polan's analysis that a significant positive association existed between dominance value and time spent eating concentrate ($r = 0.60$, $P < 0.01$). Whether subordinate cows consume as much feed as dominants or not, they may be less efficient in its use, as Bouissou (in Signoret and Bouissou, 1971) found that adrenal weights of bottom-ranking cows averaged about 20% heavier per unit of body weight than those of top-ranking cows in each of six lots, suggesting the presence of stress in those of low status.

Dairy cows are often separated into groups according to stage of lactation or level of production so that they may be fed according to their needs. This management scheme requires that animals be moved from group to group every few months. Although moving is usually associated with brief periods of increased agonistic activity, there appears to be only a temporary loss of production in carefully managed experimental herds (Arave and Albright, 1976; Brakel and Leis, 1976; Clark et al., 1977). Brantas (cited by Wieckert, 1971) found that when 15 Dutch-Friesian cows, most of which were strangers to each other, were placed together in a small area, there was intense and nearly constant activity. Only after 24 hours were several cows resting, and not until the third day were all resting simultaneously. Although the addition of strange dairy cows to a herd may well have less adverse effects than the mixing of beef cows did in the study of Schake and Riggs (1972), it is still not certain whether strange animals introduced one or a few at a time might not be stressed considerably. However, the greater availability of feed for lactating dairy cows may reduce the stress on them as compared to beef cows competing for limited feed in an otherwise similar situation.

Dairy researchers are currently evaluating the possibility of using magnetic or electronic devices attached to individual cows, which cause the release of the quantity of feed required for each individual. The feed is delivered in a stall that the cow enters temporarily and where she is undisturbed by herdmates while feeding. Such devices, currently being developed and marketed, would allow cows to be kept in more stable, socially organized groups.

Swine

Social-status effects develop early and are important in pigs. Smaller baby pigs and those born late in the farrowing period are unlikely to gain access to a high-yielding nipple (Chapter 12). After mixing of strangers (as at weaning), social orders develop rapidly, with vigorous agonistic activity. Newcomers, added a week later, are attacked and may be severely harassed. McBride and James (1964) observed that competition at feeding and watering places was pronounced. A dominant pig is relatively intolerant while feeding and often does not allow a subordinate to feed nearby, even if adequate space for two is available. McBride and James found that pigs in the lower half of the dominance order gained 7 pounds less than those in the top half from 8 to 16 weeks of age. Whether feeding lower-status pigs separately would increase their gains is not known.

Chickens

Social status and time spent feeding or near the feeder is positively correlated for hens in floor pens, as was shown by Guhl (1953) and Tindell and Craig (1959). A later study of hens in crowded multiple-hen cages by Al-Rawi et al. (1976) failed to show such an association. Why should this difference occur? Observations indicated that in roomy floor pens, where hens move about easily, dominant birds often "patrol" the feeding area and drive subordinates away. Guhl found that low-status birds tended to feed earlier in the morning and later at dusk, while their more dominant penmates were on the roost. During the day low-status hens were more likely to be seen on the roost or avoiding contact by staying out of the way. In multiple-hen or colony cages there is little room to move about under high-density conditions, and hens feed by reaching through wire openings to the feeding trough. It seems reasonable to speculate that low status does not interfere so much with feeding in cages because subordinate birds are relatively protected while feeding.

VICES RELATED TO
FEEDING-LIKE BEHAVIOR

A "vice" is an abnormal behavior that is detrimental to the health or usefulness of an animal or to others with which it is associated. A number of vices are typically associated with keeping animals in either solitary confinement or intensive production systems and include pecking, biting, sucking, or chewing on nonfood items or other animals. Aggressive acts and vices involving pecking or biting other animals are usually considered as different activities. To aid in maintaining the distinction, it may be kept in mind that aggressive acts are usually directed at the head, neck, or forequarters, at least initially. Vices associated with pecking, biting, or sucking usually involve other parts of the body of other animals and may appear initially as grooming, play, investigatory, or as the appetitive phase of feeding behavior. Vices also include acts directed at the animal's own body, as when calves or cows suck themselves and when dogs and horses engage in self-mutilation. Once established, such behaviors may occur repeatedly and can have serious or even fatal results.

Dogs and Cats: Chewing and Scratching

Biting, chewing, and manipulating nonedible objects with the forepaws by kittens and puppies is at first a means of investigating the unknown, and if the objects are found to be harmless, later involve play. These are behaviors apparently shared with all higher animals, at least among the young, and they aid the learning process.

Cats and dogs, particularly those left alone, may become destructive of furniture, clothing, and the like. Those chewing and scratching activities are not usually harmful to the animal itself but can be a great nuisance to the human beings who live in the same house. Such activities appear to be similar to those of active species of zoo animals. They may serve as "safety valves," allowing animals to cope better with confinement and loneliness. It would be better not to leave them alone for long periods of time and to avoid close confinement of active animals if possible; a spacious outdoor pen with one or more objects available for play, scratching, or chewing would be better. The best solution would be to provide ample opportunity for exercise of the type for which the animal is physically adapted. If animals must be confined for long periods inside, it would be better if they were kept in a room with minimal furniture and provided with play objects.

Horses: Wood Chewing and Pica

The eating of unnatural food is referred to as "pica." It occurs in several species, including horses kept in stalls and fed highly concentrated rations. When wood is available under such conditions, it may be chewed away in large quantities and feces may also be eaten (coprophagia). Haenlein et al. (1966) found that ponies being fed pelleted hay engaged in so much wood chewing that their wooden mangers were progressively reduced in size. Nevertheless, pelleted diets are likely to continue being fed by some owners because of economy of labor and ease of transportation, storage, and feeding.

The effect of altering diet and digestive tract acidity on feeding behavior of horses has been investigated by Willard et al. (1977), who kept their animals in metal stalls with concrete floors covered by rubber mats. A board was attached to the stall to determine to what extent the feeding of hay contrasted to a concentrated diet (with or without a chemical modifying acidity) would have on wood chewing. Electrically timed motion pictures were taken (5 seconds every 5 minutes) to provide information on behavior. Horses fed hay spent 40% of their time feeding as contrasted to less than 4% for those receiving concentrates. Although the horses chewed wood on all diets, they had five times as much of this activity when fed concentrates and chewed up more than 1 kg (2.2 lb) daily. When receiving concentrated feed, they spent nearly 3% of their time chewing feces and had more "searching" and standing time than when fed hay. Circumstantial evidence suggests that boredom caused by close confinement and the brief time required to satisfy nutritional needs when fed concentrates is an important cause of feeding vices in horses.

Chickens: Cannibalism and "Feather Pecking"

Cannibalistic pecking is directed toward blood, bleeding tissue (skin or muscle), or internal organs and occurs in both uncrowded floor pens and multiple-hen cages. It usually resembles feeding behavior and when several hens are attracted to an injured bird, the results can be deadly within a day or two. The vent or cloacal area is particularly vulnerable for hens kept in cages without nest areas as the uterus is everted during egg laying and is an attractive target. However, pecking of other areas can also be fatal; the tail region is frequently involved and areas where feathers have been lost, so that bare skin is exposed and may be scratched, causing bleeding to begin. Allen and Perry (1975) found that much of the cannibalism in their adult hens occurred before feather loss was very advanced. Therefore, it appears that although heavy feather loss may make birds more susceptible to the vice, it is not a necessary condition.

Preventive beak trimming is widely used in many countries to reduce cannibalism, but that practice is now officially discouraged in some European countries, unless an outbreak is already under way (as in Great Britain; MAFF, 1971). Even when birds are inspected frequently and injuries are treated promptly, deaths from cannibalistic pecking can occur with high frequency in the absence of beak trimming. Thus, Allen and Perry (1975) lost 3.7% of their pullets because of cannibalism during an 18-week growing period and an additional 10.6% during the subsequent year of laying in multiple-hen cages. However, the unpredictability of the situation is indicated by the same researchers, who had not previously encountered the problem although using a flock of the same genetic stock under apparently identical management in the same chicken house.

The term "feather pecking" was adopted by Hughes and Duncan (1972) in describing feather loss and hemorrhaging of skin seen in chickens. This terminology has also been used by others, but it implies a single cause of feather loss and injury to skin, which is probably an oversimplification. Feather pecking itself was not the measurement used, but rather a score based on the amount of feather damage, feather loss, or hemorrhaging of broken skin. Certainly, feather pecking and pulling of feathers is seen, but injuries from toenails, from trampling by others, and possibly by forceful collision with objects may also cause such damage. Martin et al. (1976 and personal communication) found that removal of toenails was not only associated with better feather coverage of hens, but also with reduced mortality, higher rate of lay, and better feed conversion in seven-bird cages. Hansen (1976) found that pain associated with broken feathers and torn skin appeared to be responsible for outbreaks of hysteria. When he removed toenails of day-old chicks, he hound that hysteria did not occur in cages with high "population pressure," although it did in 9 of 12 flocks with intact toenails.

Feather loss may not only cause pain and result in exposure of skin to possible injury, but may also be of direct economic significance. Hens kept in nonheated houses in cold climates will undoubtedly have higher feed requirements in order to maintain their body temperature if they lose their insulated coat of feathers.

Hormonal Effects. Sex-related hormones have major effects on behavior. Hughes (1973) investigated the effects of some hormones on feather loss and skin damage of adolescent and young-adult females. Untreated pullets showed increasing losses of feathers from 3 to 12 weeks of age and again after 18 weeks old. Hormone pellets were implanted in experimental groups at 12 weeks of age; dosages were characterized as being small, but they had large effects on feathering loss of females kept in five-bird cages. Thus, testosterone was effective in keeping feather damage and loss lower than in the untreated hens. Although estrogen and progesterone had minimal effects when given separately, they caused rapid loss of feathers when given together (i.e., they acted synergistically). Hughes noted the interesting fact that although testosterone is known to increase aggressive pecking, it had a beneficial effect on feathering scores.

Increased feather loss, vent pecking, and cannibalism are likely to occur at about the time when pullets begin to lay (Hughes and Duncan, 1972; Hughes, 1973; Allen and Perry, 1975). From the Hughes study it appears as if the combined estrogen–progesterone treatment brought on increased feather loss and vent pecking soon after begun at 12 weeks of age, even though the pullets were not ready to start laying at that age. The evidence indicates that internal hormonal environment indirectly influences feather loss and cannibalism.

External Environmental Effects. Feather loss, skin damage, and cannibalism are of particular concern in those countries where legislation has been enacted that restricts preventive measures such as beak trimming and toenail removal until a serious problem has already been demonstrated to exist. However, even beak trimming and declawing do not completely guarantee absence of feather loss and cannibalism.

External environmental conditions "causing" behavioral problems can be complex. Probably the major cause of this complexity is in the existence of "interactions." As an example, Hughes and Duncan (1972) found that one strain of chickens showed only slightly greater feather damage in cages of eight than in cages of four, while another strain had much greater damage in the larger-size cages. If they had tested the first strain only, they would have concluded that group size had essentially no effect, but if using the second strain only, they would have reached the opposite conclusion. If a number of conditions making feather damage and cannibalism more likely occur together, the probability that those vices will be experienced becomes high.

The genetic variable appears to be an important one. Therefore, it would appear relatively simple for the poultry breeder to go a long way toward solving these problems. One approach would be for the breeder to select against feather loss and cannibalism under those adverse conditions likely to produce them in susceptible stocks. There may be a serious flaw in this approach. Because other birds probably play a large role in actually causing the problems, elimination of those susceptible to feather loss and cannibalism may in fact favor selection of those individuals that are causing the problems. However, a more sophisticated selection procedure could be used.

Thus, if particular multiple-hen cages were filled with pullets all belonging to the same family and those cages were kept under conditions likely to cause the vices, then selection of entire families doing well under those conditions should give a response in the right direction. But what if genotypes that favor calm dispositions with low propensity for feather and cannibalistic pecking also result in poor productivity? Clearly, more experimental evidence must be gathered as to the likely overall success of concurrent selection for increased productivity and decreased behavioral vices before commercial breeders are likely to attempt it.

A brief list of some environmental factors that have been demonstrated to cause feather damage and loss or cannibalism in susceptible genetic stocks is given in Table 13-4. Other possible causes were reviewed by Hughes and Duncan (1972), but those need further testing. As may be seen from Table 13-4, pelleted feed, bright lights, and keeping birds in cages at high density and in larger group sizes are all practices that may cause these behavioral vices, especially when beak trimming and toenail clipping (in cages) are not done and a genetically susceptible strain is used.

Swine: Tail Biting

Comfortable pigs spend a considerable amount of time sleeping or in the drowsy state; Ruckebusch (1972) indicated a total of about 13 hours daily and Van Putten (1969) suggested 80% of the day. However, in a "barren" pen, especially in the absence of straw, pigs may become restless and are likely to bite and chew at a variety of objects, including other pigs' ears and tails. Bites to the ears invite retaliation, but tails are a safer target. Van Putten (1969) observed

Table 13-4   *Some environmental factors that increase the probability of occurrence of feather damage or loss or of cannibalism in poultry*

Causative factors	Coded reference[a]
Pelleted feed (rather than mash)	B, S&P
Bright lights	A&P, H&D
Cage housing (multiple-bird cages)	H&D
Larger group sizes in cages	A, A&P, H&D, H&H
Higher densities in cages	H&H

[a]References: A, Adams et al. (1978); A&P, Allen and Perry (1975); B, Bearse et al. (1949); H&D, Hughes and Duncan (1972); H&H, Hill and Hunt (1978); S&P, Skoglund and Palmer (1961).

tail-biting behavior closely; he believes that playful mouthing of the tail causes an accidental bite with minor bleeding which then hurts enough so that the wounded pig slashes its tail about vigorously. This draws the attention of and further biting of the bloody tail by other pigs. The pig under attack may be chased about and its tail bitten violently until only a bleeding stump is left. Van Putten lists possible consequences of tail biting as (1) restlessness and poor growth, (2) possible death during tail-biting episodes, (3) paralysis and death caused by infections, and (4) rejection of the carcass during inspection at the slaughterhouse.

Few experimental studies of tail biting have been carried out, even though Gadd (1967) stated that it is one of the most common problems in young pigs. On the basis of his extensive survey, Gadd suggested that nutrition seemed to be involved in about two-thirds of all cases and that high-energy, low-fiber, vegetable protein diets were particularly suspect. In response to this survey, Ewbank (1973) attempted to induce tail biting by feeding the kind of diet indicated. Straw was present in some pens but not in others, and some pens were poorly ventilated on a temporary basis (by accident), but Ewbank found no tail biting in any of five pens of pigs fed the special ration or in five other pens fed a "control" ration. Obviously, this behavioral problem is a difficult one to deal with experimentally.

Many producers routinely dock the tails of their baby pigs, hoping thereby to prevent the problem of tail biting. Nevertheless, a stump remains that may be bitten, and although docking is believed to be helpful as a preventative measure, reliable data on its effectiveness are not available. The practice of docking is not recommended in Great Britain except when prescribed by a veterinary surgeon (MAFF, 1971).

REFERENCES

ADAMS, A. W., J. V. CRAIG, and A. L. BHAGWAT, 1978. Effects of flock size, age at housing, and mating experience on two strains of egg-type chickens in colony cages. Poult. Sci. 57:48–53.

ALLEN, J., and G. C. PERRY, 1975. Feather pecking and cannibalism in a caged layer flock. Br. Poult. Sci. 16:441–451.

AL-RAWI, B. A., J. V. CRAIG, and A. W. ADAMS, 1976. Agonistic behavior and egg production of caged layers: genetic strain and group-size effects. Poult. Sci. 55:796–807.

ARAVE, C. W., and J. L. ALBRIGHT, 1976. Social rank and physiological traits of dairy cows as influenced by changing group membership. J. Dairy Sci. 59:974–981.

ARNOLD, G. W., and R. A. MALLER, 1974. Some aspects of competition between sheep for supplementary feed. Anim. Prod. 19:309–319.

——, 1977. Effects of nutritional experience in early and adult life on the performance and dietary habits of sheep. Appl. Anim. Ethol. 3:5–26.

BEARSE, G. E., L. R. BERG, C. F. McCLARY, and V. L. MILLER, 1949. The effect of pelleting chicken rations on the incidence of cannibalism. Poult. Sci. 28:756.

BEILHARZ, R. G., 1979. Competitive order as a measure of social dominance in dairy cattle: a criticism of the paper by Friend and Polan. Appl. Anim. Ethol. 5:191–192.

BOSC, M. J., M. F. BOUISSOU, and J. P. SIGNORET, 1968. Conséquences de la hiérarchie sociale sur le comportement alimentaire des bovins domestiques. 93rd Congrès National des Sociétés Savantes. Tours. Sciences 2:511–515.

BOUISSOU, M. F., 1970. Role du contact physique dans la manifestation des relations hiérarchiques chez les bovins. Conséquences pratiques. Ann. Zootech. 19:279–285.

——, 1971. Effet de l'absence d'informations optiques et de contact physique sur la manifestation de relations hiérarchiques chez les bovins domestiques. Ann. Biol. Anim. Biochem. Biophys. 11:191–198.

BRAKEL, W. J., and R. A. LEIS, 1976. Impact of social disorganization on behavior, mild yield, and body weight of dairy cows. J. Dairy Sci. 59:716–721.

CLARK, P. W., R. E. RICKETTS, and G. F. KRAUSE, 1977. Effect on milk yield of moving cows from group to group. J. Dairy Sci. 60:769–772.

EWBANK, R., 1973. Abnormal behaviour and pig nutrition. An unsuccessful attempt to induce tail biting by feeding a high energy, low fibre vegetable protein ration. Br. Vet. J. 129:366–369.

FRIEND, T. H., and C. E. POLAN, 1978. Competitive order as a measure of social dominance in dairy cattle. Appl. Anim. Ethol. 4:61–70.

——, and M. L. McGILLIARD, 1977. Free stall and feed bunk requirements relative to behavior, production and individual intake in dairy cows. J. Dairy Sci. 60:108–116.

GADD, J., 1967. Tail-biting: causes analyzed in 430-case studies. Pig Farming 15(7):57–58.

GUHL, A. M., 1953. Social behavior of domestic fowl. Kans. Agric. Exp. Sta. Tech. Bull. 73.

HAENLEIN, G. F. W., R. D. HOLDREN, and Y. M. YOON, 1966. Comparative response of horses and sheep to different physical forms of alfalfa hay. J. Anim. Sci. 25:740–743.

HANSEN, R. S., 1976. Nervousness and hysteria of mature female chickens. Poult. Sci. 55:531–543.

HILL, A. T., and J. R. HUNT, 1978. Layer cage depth effects on nervousness, feathering, shell breakage, performance, and net egg returns. Poult. Sci. 57: 1204–1216.

HUGHES, B. O., 1971. Allelomimetic feeding in the domestic fowl. Br. Poult. Sci. 12:359–366.

——, 1973. The effect of implanted gonadal hormones on feather pecking and cannibalism in pullets. Br. Poult. Sci. 14:341–348.

——, and I. J. H. DUNCAN, 1972. The influence of strain and environmental factors upon feather pecking and cannibalism in fowls. Br. Poult. Sci. 13: 525–547.

KATZ, D. and G. REVESZ, 1921. Experimentelle Studien zur vergleichenden Psychologie (Versuche mit Huhnern). Z. Angew. Psychol. 18:307–320.

KIDWELL, J. F., V. R. BOHMAN, and J. E. HUNTER, 1954. Individual and group feeding of experimental beef cattle as influenced by hay maturity. J. Anim. Sci. 13:543–547.

KILGOUR, R., 1978. The application of animal behavior and the humane care of farm animals. J. Anim. Sci. 46:1478–1486.

LYNCH, J. J., 1980. Behaviour of livestock in relation to their productivity. *In* M. Rechcigl, Jr. (Ed.), CRC Handbook Series in Nutrition and Food. (In press.) CRC Press, Inc., West Palm Beach, Fla.

M.A.F.F., 1971. Codes of recommendations for the welfare of livestock. Code No. 1, Cattle. Code No. 2, Pigs. Code No. 3, Domestic fowls. Code No. 4, Turkeys. Code No. 5, Sheep. Ministry of Agriculture, Fisheries and Food, London.

MARTIN, G. A., J. R. WEST, G. W. MORGAN, and T. R. BURLESON, 1976. Effects of wing and toe amputation on layers. Poult. Sci. 55:2061 (Abstract).

McBRIDE, G., and J. W. JAMES, 1964. Social behavior of domestic animals. IV. Growing pigs. Anim. Prod. 6:129–139.

McPHEE, C. P., G. McBRIDE, and J. W. JAMES, 1964. Social behaviour of domestic animals. III. Steers in small yards. Anim. Prod. 6:9–15.

NOLAN, J. V., B. W. NORTON, R. M. MURRAY, F. M. BALL, F. B. ROSEBY, W. ROHAN-JONES, M. K. HILL, and R. A. LENG, 1975. Body weight and wool production in grazing sheep given access to a supplement of urea and molasses: intake of supplement/response relationship. J. Agric. Sci., Camb. 84:39–48.

RUCKEBUSCH, Y., 1972. The relevance of drowsiness in the circadian cycle of farm animals. Anim. Behav. 20:637-643.

SCHAKE, L. M., and J. K. RIGGS, 1972. Behavior of beef cattle in confinement. Tex. Agric. Exp. Sta. Tech. Rep. 27.

SIGNORET, J. P., and M. F. BOUISSOU, 1971. Adaptation de l'animal aux grandes unités. Etudes de comportement. Bulletin Technique d'Information 258:367-372. Ministère de l'Agriculture, France.

SKOGLUND, W. C., and D. H. PALMER, 1961. Light intensity studies with broilers. Poult. Sci. 40:1458.

TINDELL, D., and J. V. CRAIG, 1959. Effects of social competition on laying house performance in the chicken. Poult. Sci. 38:95-105.

VAN PUTTEN, G., 1969. An investigation into tail-biting among fattening pigs. Br. Vet. J. 125:511-516.

WAGNON, K. A., 1965. Social dominance in range cows and its effect on supplemental feeding. Calif. Agric. Exp. Sta. Bull. 819.

——, R. G. LOY, W. C. ROLLINS, and F. D. CARROLL, 1966. Social dominance in a herd of Angus, Hereford, and Shorthorn cows. Anim. Behav. 14:474-479.

WARNICK, V. D., C. W. ARAVE, and C. H. MICKELSEN, 1977. Effects of group, individual, and isolation rearing of calves in weight gain and behavior. J. Dairy Sci. 60:947-953.

WIECKERT, D. A., 1971. Social behavior in farm animals. J. Anim. Sci. 32:1274-1277.

WILLARD, J. G., J. C. WILLARD, S. A. WOLFRAM, and J. P. BAKER, 1977. Effect of diet on cecal pH and feeding behavior of horses. J. Anim. Sci. 45:87-93.

14

Behavioral and Psychological Stress

Although the words "stress" and "distress" are somewhat ambiguous and may have somewhat different meanings for different people, they are generally understood as indicating an unfavorable state of being brought on by unpleasant or harmful conditions or "stressors." Certain hormones show typical response patterns and cause symptomatic physiological changes in many situations that we consider as stressful. When those hormonal patterns or their physiological consequences are detected, it may reasonably be expected that one or more stressors are responsible.

A complication in the study of stress is that considerable variation exists among individuals, between the sexes, and between genetic stocks. Thus, among male Rhesus monkeys, hormones known to increase following stress also differed *before* stress so that animals with highest baseline values had two to three times more than those with lowest values (Mason et al., 1968a; Mason et al., 1968b). The same kind of differences were recorded during stress, even though all animals responded to some degree. Sex differences in response to social grouping of strangers are very large in laboratory mice (Bronson, 1967), presumably because males fight vigorously, but females do not. Genetic differences in response to stressors can also be profound, as shown in a bidirectional selection study with turkeys (Brown and Nestor, 1973, 1974). Because of such studies it is clear that samples taken from animals for which baseline values have not been established will give measurements that are difficult to interpret in terms of whether they are under stress.

In working with domestic animals it is important to understand what situations are likely to cause stress, what happens when stressors upset the body's homeostatic state, and what can be done to alleviate harmful effects if stressors cannot be avoided.

HORMONAL INDICATORS

Cannon's Fight-or-Flight Response

An American physiologist, Walter Cannon, was responsible for identifying a hormone secreted by the adrenal glands in response to physiological stimuli which prepares and supports the body in *fight-or-flight* situations. Along with de la Paz, he found that cats, placed in a holder and frightened by a barking dog, secreted "adrenalin" (Cannon and de la Paz, 1911). Cannon later found that other emotional situations, such as taking difficult examinations or being involved in exciting football games, even as a coach or a player sitting on the bench, often were associated with adrenaline increases. Others soon determined that it was the inner or medullary part of the adrenal gland which produced adrenaline (reviewed by Mason, 1968b).

In the mid-1940s the hormone noradrenaline was identified and its secretion by the medulla of the adrenal gland was established. The two hormones adrenaline and noradrenaline (also called epinephrine and norepinephrine, respectively) are produced by different cell types, and stimulation of different hypothalamic areas in the cat's brain causes the selective release of one or the other. Thus, they have been shown to be independently released from the adrenal medulla following nervous stimulation arising in the hypothalamus. Research by Mason's group (Mason, 1968b) showed that secretion of high levels of those hormones may continue for periods of at least several days and that adrenaline may be primarily involved in high anxiety situations and noradrenaline more in the recovery phase following such episodes.

Selye's Stress Syndrome

The general concept of *stress* was developed by Hans Selye (1950). As a medical student in Prague, he was impressed by the "syndrome of just being sick" (Selye, 1976). Beginning with research at McGill University in 1935, he found that injecting various tissue and glandular extracts produced a triad of symptoms in laboratory rats. Those symptoms were (1) adrenal cortex enlargement, (2) lymphatic tissue atrophy, and (3) bleeding ulcers in the stomach and duodenum. After several experiments he began to be concerned that his results were caused by nonspecific tissue damage. To check this idea, the noxious substance formalin was injected and the triad of symptoms occurred within 48 hours. Other agents were also found to act as stressors and to produce the same triad of nonspecific responses. Excessive cold or heat, x-rays, excessive exercise, and injury were all implicated. Emotional upset, behavioral threats, and other psychological influences are also now recognized as potent stressors, even if direct physical injury is not involved.

When stressors continue to act over a prolonged period and are not so extreme as to be fatal in the short run, then Selye described a *general adaptation syndrome* as occurring. This syndrome was visualized as having three distinct phases:

1. *Alarm reaction*—The triad of symptoms appears and "resistance" to more extreme stress of the same kind or to other kinds of stress is reduced.

2. *Stage of resistance*—The animal adapts for a period of time and appears to have overcome the stress. During this interval it may be subjected to an increase of stress of the same kind with fewer ill effects than would occur in a previously unstressed animal.

3. *Stage of exhaustion*—The animal may again show one or more of the triad of symptoms; its resistance drops under continuing action of the stressor and it becomes obviously "sick" and may die.

Selye believed that most animals (including man) go through the first two stages many times in a normal life span; only long-lasting stressors would produce the third stage. He showed that stressors have additive effects. Thus, if each of two animals are stressed, one by cold, the other by injury, then each may have symptoms such as moderate adrenal cortex enlargement. However, if both stressors act on a single animal at the same time, the symptoms become much more evident (e.g., the adrenal cortex may become greatly enlarged). When the adrenal cortex responds it is because the anterior pituitary gland has increased its output of adrenocorticotropic hormone (ACTH).

When a stressor acts on an animal, messages are received by the hypothalamic area of the brain. Thus alerted, the hypothalamus responds in two basic ways, as shown in Figure 14-1. Fast-traveling nerve impulses pass along the sympathetic nervous system to the adrenal medulla, as previously discussed, which then increases secretion and synthesis of adrenaline, and in some cases noradrenaline. The other hypothalamic response involves hormonal pathways which usually require minutes or hours before measurable responses are apparent, in contrast to the nearly instantaneous fight-or-flight responses. A "releasing factor" passes from the hypothalamus into the anterior pituitary, thereby increasing ACTH secretion. In response to

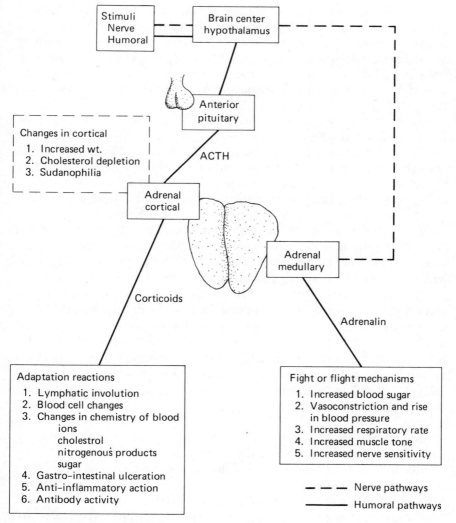

Figure 14-1 Schematic summary of systemic changes associated with stress. (From Siegel, 1971.)

ACTH stimulation, the adrenal cortex secretes steroid hormones or "corticoids," which have a variety of physiological effects; some of which are indicated in Figure 14-1.

Selye proposed that under ordinary circumstances in the unstressed animal, pro-inflammatory hormones produced by the adrenal cortex have the upper hand over anti-inflammatory hormones, thereby keeping the animal in readiness to respond to local injuries. If an animal is wounded, the local area becomes inflamed by dilation of blood vessels and, among other effects, white blood cells rush to the scene, where certain types engulf and digest pathogenic organisms. Ironically, pro-inflammatory hormones sometimes are involved in malfunctioning, when they cause the body to overreact. Thus, hay fever is an inflammatory response in the lungs brought on by pollen that is otherwise harmless, and arthritis is associated with inflammatory responses in the joints.

As a stressor continues to act, anti-inflammatory hormones such as cortisol, cortisone, and corticosterone produced by the adrenal cortex gain ascendency over the pro-inflammatory hormones. There are then rather drastic alterations in body chemistry. The second line of defense, the "adaptation reactions," come into play. Among the changes, we find that (1) inflammation is reduced if a physical injury has occurred or if tissues are inflamed; (2) the metabolic rate increases, protein and fatty tissues break down, and blood sugar rises; (3) the lymphatic organs (thymus, spleen, etc.) atrophy and the total white blood cell count decreases, and the proportion of different leukocyte types is altered; (4) antibody production decreases, (5) reproductive capacity decreases; and (6) ulceration of the stomach and intestinal tract may occur (in mammals, but not in birds).

That psychological influences, as well as physical stressors, could produce the stress syndrome was recognized by Selye (1950), who wrote: "Even mere emotional stress, for instance that caused by immobilizing an animal on a board (taking care to avoid any physical injury), proved to be a suitable routine procedure for the production of a severe alarm reaction."

Social Stress. A variety of deleterious symptoms are typically associated with crowding, which reduces the fitness of the population involved. Christian (1955) found that adrenal weights of male mice increased as density increased in groups of constant size. Similar indications of increased adrenal activity were later found in wild rats, wild deer, and monkeys when crowded or grouped (reviewed by Mason, 1968a).

Since the 1950s poultry have been crowded more and more;

chickens, in particular, have been confined inside, and area per bird has been reduced again and again as labor-saving devices were installed and housing costs increased. Poultry producers were concerned for the welfare of their birds, especially as it affected productivity, and scientists began looking for indications of stress. Siegel (1959) examined the effects on laying hens. Although all his hens had the same amount of roosting, feeding, nesting, and watering space per individual, those kept in flocks of 150 with 1.33 ft² (0.12 m²) of area for each hen had heavier adrenal gland weights associated with an apparently abnormal increase in the cortical area when compared with those kept in flocks of 50 with three times as much floor space. Not surprisingly, hens in the more crowded flocks laid fewer eggs. Nevertheless, Siegel concluded that "at the floor space levels tested the stress imposed was sub-acute in nature, probably well within the birds' adaptive abilities."

Studies by Siegel and Siegel (1961) involved cockerels in various social situations. Left adrenal gland weights were found to be fairly sensitive indicators of the relative stressfulness of those situations. Because of differences in ages and genetic sources of cockerels in the different experiments, the results need to be compared with some caution. Nevertheless, the following tentative conclusions appear to be reasonable:

1. *Ability to win contests.* The relative ability to win pair contests with strangers was not associated with adrenal response. This was a minimal kind of stress, probably because individuals were separated as soon as a dominance relationship was determined.

2. *Low status.* Low social standing in undisturbed groups was of intermediate stressfulness. Low-ranking individuals had heavier adrenals in some groups but not in others.

3. *Physical isolation vs. grouping.* Cockerels kept in individual cages that prevented physical contact but allowed them to see and hear others had lighter-weight adrenals than those of males kept continuously in groups in floor pens.

4. *Daily exposure of individuals to groups of strangers.* Individual males moved daily from small organized groups into pens of strange cockerels for 4-hour periods were attacked vigorously each time and spent most of the period attempting to hide under feed hoppers or in corners. After a 22-day experimental period, they had heavier adrenals than males kept continuously in small organized groups during the same period.

Some investigators (e.g., Southwick and Bland, 1959) were inclined to believe that adrenal gland enlargement in crowded groups, especially when males were involved, resulted primarily from physical injuries rather than from psychological stressors. A study by Bronson and Eleftheriou (1965) showed how the two kinds of effects, physical vs. psychological, could be separated. They were able to use a sophisticated assay to reflect adrenal cortex activity. Three groups of male mice were used, which had been kept in isolation until tested. Individual mice of the experimental groups were placed in a cage with a highly aggressive "fighter" mouse for 15 minutes daily for at least 5 days. Males of one group were always separated from the fighter by a partition (control group); males of another group always had the partition removed during the last 5 minutes and were attacked and subjugated daily; third-group individuals were also subjected to attacks by the fighter (in the same way) on each of five successive days, but were subsequently separated by the partition for the total period, for up to 4 more days. Males of the control group (not attacked) had relatively low values of unbound corticosterone, those that were physically attacked daily had higher readings, and those that were initially attacked but were later protected from physical contact had the highest values (although not significantly more than the second group). Because the hormone rose to its peak concentration about an hour after exposure for males attacked by a fighter and because concentrations returned to trace amounts by 3 hours after each exposure, it was clear that there was no carryover from treatment on previous days. Therefore, Bronson and Eleftheriou had demonstrated that the psychological stressor in their study was at least as powerful as direct physical attacks in causing adrenal cortex response.

Psychological vs. Other Stressors. Mason and his coworkers at the Walter Reed Army Medical Center in Washington, D.C., carried out numerous studies on Rhesus monkeys and man that have stimulated renewed interest in psychological stress. Just as Bronson and Eleftheriou were able to separate the effects of psychological and physical stressors in mice, Mason's group found that some presumed environmental stressors either have no detectable effect or unexpected effects if presented in ways that are not psychologically disturbing.

Mason (1971) describes how in studying the effects of fasting, two monkeys of a group of eight were suddenly deprived of food for 3 days while their neighbors continued eating as usual. The sights, sounds, and odors associated with normal feeding were present while those two were ignored when food was passed out. The deprived monkeys protested loudly but without effect. It was assumed that the discomfort of an empty stomach added to their psychological reaction. In this situation the fasting monkeys showed a marked increase in urinary corticosteroid. Mason's group then attempted to minimize the psychological variables. This was done by placing fasting monkeys in small, private cubicles and feeding them fruit-flavored but nonnutritive pellets similar in taste and appearance to those used in their normal diet. In this situation there was no significant increase in the urinary corticosteroid, although changes in epinephrine, norepinephrine, insulin, and growth hormone occurred.

On the basis of studies with seven presumed stressors, Mason (1975) argues that increased levels of adrenal corticoids can generally be used to indicate when animals are under psychological stress, but physical and physiological stressors may, or may not, have such an effect.

Optimal Levels of Stress in Early Life? Although symptoms of stress are generally regarded as indicating that all is not well, studies with laboratory rats suggest that relatively brief periods of exposure to stressors during the infantile period elicit adrenal response and may be beneficial in preparing the animal for challenges encountered later in life. Thus, 2-day-old rats exposed to various stressors exhibited elevated levels of corticosterone (e.g., Zarrow et al., 1967) and rats "handled" in infancy respond more promptly to an acute stressor (electric shock) as adults. Levine (1962) found that adult rats which had been handled in infancy had significantly elevated levels of circulating corticosteroid within 15 seconds following the onset of intermittent shocks, but nonhandled controls did not show a clear response until 5 minutes had passed. In survival tests, rats handled from birth until weaning survived longer when totally deprived of food and water than did controls (Levine and Otis, 1958; Denenberg and Karas, 1959).

From the studies with laboratory rats, it might be assumed that a certain level of infantile stress would be beneficial in other species as well. Experiments with mice indicate, however, that what is optimal for one species may not be for another. In the same report dealing with rats, Denenberg and Karas (1959) also presented results for mice handled in the same way (i.e., with handling for either 10 days or 20 days or without handling before weaning). When subjected to the stress of total food and water deprivation as adults, mice handled for 20 days lived less well than controls, while those manipulated for 10 days did not differ from unhandled mice.

It has been postulated that effective behavior may depend on some optimum level of stress. Levine (1971) noted that ACTH secreted from the anterior pituitary following stress appears to sharpen the senses in man, especially the ability to taste, smell, hear, and detect internal signals from the body. It also decreases the rate of habituation to novel environmental stimuli and slows the loss of conditioned avoidance responses. All of these effects would appear to be adaptive to animals in keeping them more sensitive to the environment when stressors have activated the hypothalamus and thereby increased ACTH secretion. As glucocorticoids arise in the body (in response to ACTH), those effects are offset.

Hormone Profiles. Mason et al. (1968a) trained five healthy, adult male monkeys to avoid electric shocks during a 72-hour period. A red light remained on continuously during the conditioned avoidance sessions, and the animals could prevent shock by pressing a hand lever at intervals of less than 20 seconds. The monkeys were fairly efficient at preventing shocks; average rates of lever pressing rose from several hundred a day during the preshock period to about 30,000 a day during the period when they were required to prevent shock. During the postavoidance period, they dropped off rapidly. Multiple hormonal changes associated with the 3-day conditioned avoidance period and the following 6-day recovery period have been summarized by Mason (1968c, 1975). Those responses are shown in Figure 14-2. The adrenal glucocorticoid response is reflected by the 17-OHCS values (urinary 17-hydroxycorticosteroid hormone), which rose abruptly during the avoidance period, continued at a high level on the first postavoidance day, and then gradually dropped to near-basal levels during the next 5 days. Closer examination of the first-day response indicated that an acute elevation occurred during the first 2 hours, but this was not a maximum response, as previous studies indicated that nearly twice as much could be elicited by large doses of ACTH injection. It was assumed that the predominant stressor was the psychological disturbance associated with the possibility of being shocked. However, after the first 12 hours the lack of sleep and the work of pressing the easily moved lever may have become stressors also.

The prolonged effect of the stress situation on epinephrine (adrenaline) secretion and the rise of the other adrenal medulla hormone, norepinephrine (noradrenaline), are of special interest. As was mentioned earlier, those two hormones, especially epinephrine, have typically been thought of as preparing and sustaining the body in fight-or-flight situations (see Figure 14-1). Most other studies have followed their course for relatively brief periods, usually for a few hours at most. In Figure 14-2 we see the differential response between epinephrine and norepinephrine very clearly; epinephrine remained high during the 3-day avoidance period but dropped to basal levels immediately following, but norepinephrine rose to even higher levels during the 6-day, postavoidance period.

Besides the adrenal hormone responses usually detected as accompanying stress, Mason's group found that all other hormone systems examined also showed responses. It is apparent that the complex set of responses occurring during and following stressful behavior situations would cause many physiological changes. Mason (1968c, 1975) suggests that those changes promote mobilization of energy by the breaking down of the body's resources and prevent the building up of protein and energy stores. During the recovery period, opposite metabolic processes occur as the body attempts to restore its depleted resources and return to a more normal state.

Hormonal Indicators in Domestic Animals. Quantitative measures of adrenal hormones in the circulating blood plasma or in the urine would appear to be useful indicators of stress. However, such techniques require a high level of training, and such measures have met with only limited success in domesticated animals. Circulating hormones associated with stress show diurnal variation, individual animal differences are large, and there are formidable problems of collection. Thus, Willett and Erb (1972), working with dairy cows, found that elaborate precautions were necessary; the acts of forcefully restraining animals and drawing blood samples from them could, in themselves, cause significant rises in hormone level. Arave et al. (1977) forced changes in dominance rank by regrouping dairy cows. Assuming that such social activities were stressful, they attempted to measure the responses by hormonal changes, but concluded that "total corticoids did not appear to be a valid or useful determinant of social stress." Kilgour and DeLangen (1970) had greater success in using cortisol assays to measure stresses imposed by management practices in sheep.

Selye's triad of symptoms, found in laboratory animals under stress, suggest that measures such as adrenal weights and ulcer production may be useful in animals that are to be slaughtered or that can be sacrificed. Various supposedly stressful conditions (low social status, crowding, social disruptions, etc.) have been imposed on chickens, and increased adrenal weights have been found under some of the more extreme treatments (e.g., Siegel, 1959, 1960; Siegel and Siegel, 1961). Ulceration of the digestive tract does not occur in poul-

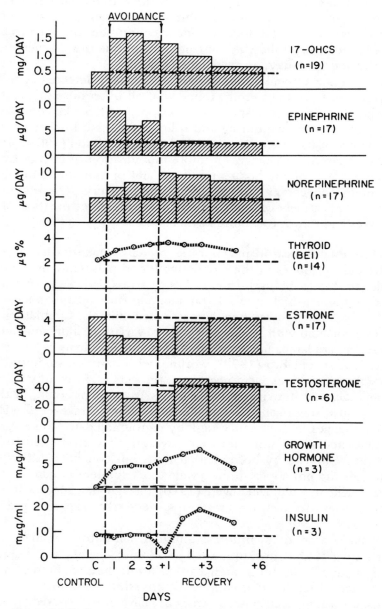

PROFILE OF HORMONAL RESPONSES TO
CONDITIONED AVOIDANCE IN MONKEYS

AVOIDANCE

17-OHCS
(n=19)

EPINEPHRINE
(n=17)

NOREPINEPHRINE
(n=17)

THYROID
(BEI)
(n=14)

ESTRONE
(n=17)

TESTOSTERONE
(n=6)

GROWTH
HORMONE
(n=3)

INSULIN
(n=3)

C   1   2   3 +1   +3   +6

CONTROL   RECOVERY

DAYS

Figure 14-2 Pattern of multiple hormonal responses in relation
to 72-hour conditioned avoidance sessions in male Rhesus monkeys; C indicates basal levels prior to the avoidance period. (From
Mason, 1968c.)

try but has been used as a criterion of stress in swine (Muggenberg et al., 1967). A careful review of various means of assaying for stress response, primarily in birds, led Siegel (1971) to conclude that it is often necessary to use several physiological criteria in evaluating an animal's total response.

BEHAVIORAL AND PERFORMANCE INDICATORS

It is usually assumed that elevated frequencies of agonistic acts within groups of animals indicate stressful social environments. Such increased frequencies typically occur when strange individuals are brought together or when a stranger is introduced into a group. Adverse effects of adding strange hens to small flocks daily or on alternate days were documented in the study of Guhl and Allee (1944). They found that disorganized flocks had higher pecking frequencies, lower feed intake, and reduced egg production when compared with undisturbed flocks. Craig et al. (1969) produced social disorganization by randomly redistributing pullets among flocks on a weekly basis, from 18 to 30 weeks of age. This treatment also increased agonistic activity, but *without significant adverse effects* on age at sexual maturity or egg production (Craig and Toth, 1969). The contrasting results of the two studies are shown in Table 14-1.

How can the differences in productivity between the two studies be accounted for? It was suggested that the social disruptions were less stressful in the Craig and Toth study because all pullets were interacting with strangers, whereas Guhl and Allee's introduced hens were attacked by relatively organized groups. Weekly redistribution of pullets may also have benefited those individuals that would otherwise have remained at the bottom of a stable peck order. Under a system of changing group membership where nearly all individuals are strangers to each other, low-status individuals have an opportunity to rise in the hierarchy whenever a new group is formed. It is apparent from the contrasting results of the two experiments and from observations of decreased aggression which occurs in some species under extremely high density (Chapter 11) that frequency of aggressive acts may be an unreliable criterion of stress.

Behaviors not seen or occurring at low frequencies in relatively natural environments but which become evident under high-density management systems are often regarded as indicating distress. Wood-Gush et al. (1975) describe a number of such activities, especially those likely to occur in frustrating situations. Chickens, in particular,

Table 14-1   *Effects of different kinds of social disorganization on relative frequency of aggressive acts and on various performance traits of hens*

Treatments			
On alternate days a single strange hen was added to each flock of seven and the longest-term resident was removed[a]		Weekly redistribution of pullets among flocks of 18, between 18 and 30 weeks of age[b]	
Effects: $\dfrac{\text{disorganized flock means}}{\text{organized flock means}} \times 100$			
Aggressive acts, frequency	157%[c]	Aggressive acts, frequency	174%[c]
Feed consumed per hen	74%[c]	Age at first egg	99%[d]
Eggs per hen	79%[c]	Rate of egg laying	96%[d]

[a]Data from Guhl and Allee (1944).

[b]Data from Craig and Toth (1969).

[c]Within-study difference between disorganized and organized flocks was greater than expected due to chance ($P < 0.05$).

[d]Within-study difference (as above) was nonsignificant (i.e., could reasonably have been caused by chance).

have been singled out for studies of abnormal behavior in highly artificial environments.

A study by Mauldin and Siegel (1979) with individually caged hens revealed that the incidence of "fear" responses to human beings and head shaking differed among genetic stocks. Fearful responses were described as modest and decreased as hens were kept in cages for longer periods. Within genetic stocks, scores for fearfulness were not associated with egg production, age at sexual maturity, or body weights. There appeared to be no important association between fear and head shaking. They speculated that head shaking and similar movements that may increase under confinement conditions may be mild responses which allow the animal to relieve tension and therefore cope better.

As in the case of frequency of aggressive acts, it appears that "abnormal" behaviors exhibited under conditions of close confinement may be poor indicators of stress, as indicated by either direct physiological measures (as for adrenal corticoids) or by indirect measures of well-being, such as body weight or reproductive functioning. Nevertheless, some abnormal behaviors, such as "cannibalism" in chickens, tail biting in pigs, "wool picking" in sheep, and excessive riding of certain steers (the "buller" syndrome) in feedlots are clearly harmful and victims are stressed.

GENETIC SELECTION FOR
STRESS RESISTANCE

Can selection for resistance to social and psychological stressors cause real changes? The veterinarians Gross and Colmano, working at Virginia Polytechnic Institute, subjected chickens to social stress by repeatedly moving them from one organized group to another. At the end of a 2-week stressing period, plasma was collected from each bird and the amount of corticosterone was determined. On the basis of this testing procedure, high- and low-response birds were selected and mated with their own kind. When progeny of the two lines were tested in the same way, they differed from each other in the same direction in which their parents were selected (Gross and Colmano, 1971). These investigators were particularly interested in how lines separated genetically on their hormonal response to social stress might differ in disease resistance. They found that the difference in plasma corticosterone response produced by genetic selection affected disease resistance; the low response line produced more antibody when challenged and was more resistant to a viral disease but was less resistant to bacterial infection. The results have opened a fascinating area for further research, but the practical implications, in terms of how behavioral stress may be alleviated, are not yet clear.

In a longer-term selection study for response to cold stress in turkeys, Brown and Nestor (1973, 1974) also found that plasma corticosterone could be significantly changed. Their high line had about twice as much plasma corticosterone as the low line (when cold stressed) after six generations of selection, and the unselected control was intermediate. It is not known how those lines would have responded to social stress, but they were found to respond in the same general way to heat stress as to cold stress.

The studies cited above indicate genetic influences on stress resistance. Clearly, some genetic stocks withstand stress better than others.

REFERENCES

ARAVE, C. W., C. H. MICKELSEN, R. C. LAMB, A. J. SVEJDA, and R. V. CANFIELD, 1977. Effects of dominance rank changes, age, and body weight on plasma corticoids of mature dairy cattle. J. Dairy Sci. 60:244–248.

BRONSON, F. H., 1967. Effects of social stimulation on adrenal and reproductive physiology of animals. In M. L. Conalty (Ed.), Husbandry of Laboratory Animals. Academic Press, Inc., New York.

——, and B. E. ELEFTHERIOU, 1965. Adrenal response to fighting in mice: separation of physical and psychological causes. Science 147:627–628.

BROWN, K. I., and K. E. NESTOR, 1973. Some physiological responses of turkeys selected for high and low adrenal response to cold stress. Poult. Sci. 52:1948–1954.

——, 1974. Implications of selection for high and low adrenal response to stress. Poult. Sci. 53:1297–1306.

CANNON, W. B., and D. de la PAZ, 1911. Emotional stimulation of adrenal secretion. Am. J. Physiol. 28:64–70.

CHRISTIAN, J. J., 1955. Effects of population size on the adrenal glands and reproductive organs of male mice in populations of fixed size. Am. J. Physiol. 182:292–300.

CRAIG, J. V., and A. TOTH, 1969. Productivity of pullets influenced by genetic selection for social dominance ability and by stability of flock membership. Poult. Sci. 48:1729–1736.

CRAIG, J. V., D. K. BISWAS, and A. M. GUHL, 1969. Agonistic behavior influenced by strangeness, crowding and heredity in female domestic fowl. Anim. Behav. 17:498–506.

DENENBERG, V. H., and G. C. KARAS, 1959. Effects of differential infantile handling upon weight gain and mortality in the rat and mouse. Science 130: 629–630.

GROSS, W. B., and G. COLMANO, 1971. Effect of infectious agents on chickens selected for plasma corticosterone response to social stress. Poult. Sci. 50: 1212–1216.

GUHL, A. M., and W. C. ALLEE, 1944. Some measurable effects of social organization in flocks of hens. Physiol. Zool. 17:320–347.

KILGOUR, R., and H. DELANGEN, 1970. Stress in sheep resulting from management practices. N.Z. Soc. Anim. Prod. 30:65–76.

LEVINE, S., 1962. Plasma-free corticosteroid response to electric shock in rats stimulated in infancy. Science 135:795–796.

——, 1971. Stress and behavior. Sci. Am. 224:26–31.

——, and L. S. OTIS, 1958. The effects of handling before and after weaning on the resistance of albino rats to later deprivation. Can. J. Psychol. 12:103–108.

MASON, J. W., 1968a. A review of psychoendocrine research on the pituitary-adrenal cortical system. Psychosom. Med. 30:576–607.

——, 1968b. A review of psychoendocrine research on the sympathetic–adrenal medullary system. Psychosom. Med. 30:631–653.

——, 1968c. Organization of the multiple endocrine responses to avoidance in the monkey. Psychosom. Med. 30:774–790.

——, 1971. A re-evaluation of the concept of "non-specificity" in stress theory. J. Psychiatr. Res. 8:323-333.

——, 1975. Emotion as reflected in patterns of endocrine integration. *In* L. Levi (Ed.), Emotions: Their Parameters and Measurement. Raven Press, New York.

——, J. V. BRADY, and G. A. TOLLIVER, 1968a. Plasma and urinary 17-hydroxycorticosteroid responses to 72-hr. avoidance sessions in the monkey. Psychosom. Med. 30:608-630.

MASON, J. W., W. W. TOLSON, J. V. BRADY, G. A. TOLLIVER, and L. I. GILMORE, 1968b. Urinary epinephrine and norepinephrine responses to 72-hr. avoidance sessions in the monkey. Psychosom. Med. 30:654-665.

MAULDIN, J. M., and P. B. SIEGEL, 1979. "Fear," head shaking and production in five populations of caged chickens. Br. Poult. Sci. 20:39-44.

MUGGENBERG, B. A., T. KOWALCZYK, W. G. HOEKSTRA, and R. H. GRUMMER, 1967. Effect of certain management variables on the incidence and severity of gastric lesions in swine. Vet. Med./Small Anim. Clin. 62:1090-1094.

SELYE, H., 1950. The Physiology and Pathology of Exposure to Stress. Acta, Inc., Montreal.

——, 1976. The Stress of Life, rev. ed. McGraw-Hill Book Company, New York.

SIEGEL, H. S., 1959. Egg production characteristics and adrenal function in White Leghorns confined at different floor space levels. Poult. Sci. 38:893-898.

——, 1960. Effect of population density on the pituitary–adrenal cortical axis of cockerels. Poult. Sci. 39:500-510.

——, 1971. Adrenals, stress and the environment. World's Poult. Sci. J. 27:327-349.

——, and P. B. SIEGEL, 1961. The relationship of social competition with endocrine weights and activity in male chickens. Anim. Behav. 9:151-158.

SOUTHWICK, C. H., and V. P. BLAND, 1959. Effect of population density on adrenal glands and reproductive organs of CFW mice. Am. J. Physiol. 197:111-114.

WILLETT, L. B., and R. E. ERB, 1972. Short term changes in plasma corticoids in dairy cattle. J. Anim. Sci. 34:103-111.

WOOD-GUSH, D. G. M., I. J. H. DUNCAN, and D. FRASER, 1975. Social stress and welfare problems in agricultural animals. *In* E. S. E. Hafez (Ed.), The Behaviour of Domestic Animals, 3rd ed. The Williams & Wilkins Company, Baltimore.

ZARROW, M. X., V. H. DENENBERG, G. C. HOLTMEYER, and J. T. BRUMAGHIM, 1967. Plasma and adrenal corticosterone levels following exposure of the two-day-old rat to various stressors. Proc. Exp. Biol. Med. 125:113-116.

15

Shelter, Welfare, and Handling

SHELTER-SEEKING BEHAVIOR

Young mammals and birds of precocial species kept in cool environments readily locate and move to sources of warmth, which may be either other animals or artificial heating devices. Overheating is avoided by moving apart or away from heat sources. As long as freedom of movement is possible and a suitable range of temperatures exists, young animals will not suffer if only a part of their environment has extreme temperatures.

Older animals have less need to seek out moderate conditions because of their better internal control of body temperatures. Given adequate feed and unfrozen water they can cope with considerable cold if their coats of feathers, hair, or wool are in good condition. Shelters are obviously needed if cold temperatures or the combined effects of temperature, humidity, and wind are beyond the animal's adaptive ability. Livestock producers will want to provide shelter against cold if increased feed intake required to keep the animal warm by metabolic processes is more costly than shelters.

Shades, ventilation, cooling sprays, water to stand or wallow in, and adequate drinking water (preferably cool) reduce heat loads when temperatures are high and by providing greater comfort allow animals to have higher feed intake, which may be desirable if they are growing or producing.

When animals are allowed little or no choice of movement in close confinement, under intensive husbandry, or when moved or restrained, more needs to be known of their requirements. Environmental variables such as temperature, humidity, air movement, light, space, and exposure to other animals may also affect activity levels, irritability, and social behavior.

Avoidance of Chilling in Sheep

Lynch and Alexander (1976) reviewed research dealing with shelter-seeking behavior of sheep during cold and windy weather. Studies were cited showing that adult, wool-covered sheep use natural or artificial shelter only erratically unless conditions are quite extreme, usually including strong winds. Because some wooly ewes do not seek adequate shelter before lambing, mortality among the newborn may be heavy. Confining sheep to shelter during lambing in inclement weather can have beneficial effects on lamb survival (Watson et al., 1968; Egan et al., 1972). Unlike wool-covered adults, growing lambs and recently shorn sheep seek out shelter readily.

Because of the findings indicated above, and because many producers may not be able to provide buildings to protect their sheep during the lambing season, Lynch and Alexander (1976) carried out a study, under New South Wales (Australia) conditions, in which a tall grass was planted in strips within a pasture to provide some shelter from the elements. The grass used was an infertile hybrid selected for vigorous growth and was relatively unpalatable, at least when mature (Figure 15-1).

Half of the ewes were shorn shortly before their expected lambing dates and all groups were placed in experimental lots (half with grass windbreaks and half without) a day before lambing was expected to begin. Following a 24-hour to 3-day period after lambing

Figure 15-1 Shorn ewes seek shelter behind grass windbreaks on cold, windy days. Lambs born behind such shelters during inclement weather survive better than if none is provided. (Courtesy of J. J. Lynch, CSIRO, Armidale, Australia.)

the ewes and their lambs were again moved to unsheltered pasture. Thus, grass windbreaks, if provided, afforded protection for a limited time only (i.e., from just before lambing until 1 to 3 days following parturition). Hourly behavioral observations were carried out in the experimental lots and the locations of ewes and their lambs were mapped. Shorn ewes spent twice as much time close to shelter belts as did unshorn ones. Shorn sheep tended to shelter when it was cold or raining and to stand on the lee side when winds were strong. Unshorn sheep tended to use the shelter for shade on warm, sunny days when there was little cooling from wind movement; they also used them one especially cold and windy 2-day period. During a 5-day period when high winds and rain prevailed, lamb mortality from exposure was significantly greater in unsheltered groups (7 of 30) than for those sheltered by high grass (1 of 41). Unshorn ewes tended to have higher lamb mortality, apparently because they remained near the grass windbreak less.

Use of Shade by Dairy Cows

High environmental temperature, high humidity, and intense solar radiation can have adverse physiological effects on dairy cows in tropical and subtropical regions which are reflected in reduced performance (reviewed by Roman-Ponce et al., 1977). Those problems may be reduced by evaporative cooling or air conditioning, but those solutions are not practical for many tropical areas. Roman-Ponce et al. looked at lactating cows' behavior and other indicators of their well-being and productivity under summertime conditions in subtropical Florida when they had access to artificial shade in one lot, but no shade in an adjacent one.

Although free to move about, cows usually remained under the shade structure during daylight hours (where feed and water were available), but moved onto an adjoining sod area in the evening and at night. Animals in the next lot, which was without either natural or artificial shade, were usually lying down in sod or wet areas during hot hours of the day, Thus, the pattern of behavior differed considerably, and the nonshaded cows preferred to eat during late afternoon and at night. Shaded cows had lower respiration rates and body temperatures, yielded nearly 11% more milk, had 19% higher conception rate, and 10% lower incidence of clinical mastitis. The improved performance for shaded cows was essentially the same as obtained with air conditioning under Florida conditions (Thatcher et al., 1974). Although some other studies involving shades have not shown the same improvement in milk production as found by

Roman-Ponce et al., it is clear that their cows sought out shade when needed and benefited from it.

Temperature-Related Behavior in Pigs

Pigs differ from most other domesticated animals in having sparse protection of the skin and a subcutaneous layer of fat for thermal insulation. The sparse hair covering allows for ready evaporation of moisture, but the pig does not sweat when exposed to heat (Ingram, 1967). Therefore, direct evaporative cooling requires external wetting of the skin or wallowing (Figure 15-2). Heat loss from the pig's skin also occurs readily when the skin is exposed to air movement. Pigs kept in cold climates can prevent considerable heat loss by huddling (Holmes and Mount, 1967). However, even an individual pig can alter its heat loss by changes in posture; a relaxed sprawling position results in greater conduction of heat away than when a pig lies on its legs (Mount, 1968).

Beginning in the 1960s, studies involving operant conditioning techniques by Baldwin and Ingram and their coworkers in Cambridge have shed much light on behavioral regulation of temperature in pigs and sheep. For example, it was learned that a pig confined in a cage at 10°C (50°F) over a period of many days would work to obtain radiant heat in every hour, but pigs housed in a hut in an outside yard behaved very differently (Ingram et al., 1975b). A bank of heaters in the yard was used by all pigs during a cold day, but at

Figure 15-2 Sows wallow in a mud hole on a hot summer day to keep their body temperature down. Swine do not sweat, but evaporative cooling from external wetting of the skin is efficient. (Photograph by the author.)

night they returned to the hut and huddled for warmth. In an earlier study, Ingram and Legge (1970) confined pigs in an area including grass paddocks, a wooded area, and a hut in the wooded area that was supplied with straw. Under those conditions the pigs did not seek shelter in the hut during the day until air temperature dropped below 5°C (about 40°F). Their pigs showed no particular tendency to seek out sunny areas or to stay out of the rain. However, they did seek shelter from cold winds; this is understandable, as it has been shown that under windy outdoor conditions heat loss may increase sixfold (Ingram et al., 1975a).

Physiology and Thermoregulatory Behavior. It is evident that laboratory studies of the regulation of heat by pigs inform us of pigs' preferences regarding temperature in that setting, but they do not indicate with much accuracy how they will behave under farm conditions. Nevertheless, laboratory studies are valuable in providing insight into physiological mechanisms influencing behavioral regulation of body temperature. Using operant conditioning methods it was learned that pigs would turn on radiant heaters in about half of all trials at temperatures of 30°C (86°F) and above when the preoptic area of the hypothalamus was cooled, although such behavior would not be shown without hypothalamic cooling (Baldwin and Ingram, 1967). However, at lower temperatures the frequency increased as ambient temperature was lowered. When the preoptic area was warmed, the frequency of turning on radiant heat was reduced even at freezing temperature. Similar studies have been carried out with sheep with comparable results (Baldwin and Yates, 1977). These experiments indicate that both peripheral and central temperature receptors influence behavior related to maintaining body temperature within acceptable limits.

PREFERENCE TESTING
AND ANIMAL WELFARE

In intensive husbandry systems animals often have little choice of environmental conditions, and there is considerable controversy as to what should be provided. Baldwin and Meese (1977) carried out a study to learn about illumination preferences. Their pigs were tested one at a time in a pen located in a soundproof room that was dark except when the pig pushed its snout through a slit in the wall, thereby interrupting an infrared beam. As long as the beam was broken, the pen was lighted. Another slit was provided which had no function other than to allow the pig to push its snout through; it served to determine whether a pig was actually seeking light by putting its nose through an opening. During trial periods, pigs kept the lights on for only 0.5% of the time. Pigs used the slit causing lights to be on significantly more than the control slit; they would work to have the lights on, but not much!

In further trials, Baldwin and Meese provided a different situation for their pigs by having interruption of one infrared beam turn lights on while interrupting another beam turned lights off. With this situation, pigs kept their lights on for 72% of the time; they tended to avoid long periods without light. We see from the two kinds of lighting preference trials that very different conclusions would be drawn if one or the other was conducted.

Welfare Interpretations

We have seen from the pig-testing situations described above how different experimental setups can lead to very different results in terms of what animals seem to prefer. Clearly, animals need to be protected from extreme conditions that cause physical injuries and unnecessary pain, or from husbandry considerations, from situations causing significant declines in productivity or increased death loss. But how far is it desirable to go? Different people have different outlooks, and some producers who might prefer to provide a more comfortable existence for their stock may be prevented from doing that by economic necessity or by lack of knowledge as to what conditions their animals would prefer.

Preconceived human notions of animals' preferences frequently prove incorrect when put to the test, or different answers are given in different tests. As an example, wire flooring for laying hens certainly is not part of a natural-type environment, and human beings would probably prefer straw or litter to a welded-wire living surface. Hughes (1976) found, in one study, that hens given constant access to both kinds of flooring did not, as a group, prefer one over the other. However, particular individuals held strong preferences for one or the other, and previous experience with either wire or litter tended to cause preference for the same flooring while on test. Even so, 88% of all eggs were laid on the litter; it was preferred as a nesting material.

Considerable controversy has arisen in some countries as to what preference testing can reveal about the "suffering" and "welfare" of animals under different systems of husbandry. Many people have been particularly concerned over the keeping of laying hens in multiple-bird, wire-floored cages, which restrict freedom of movement greatly and force hens into each other's personal space. A series of small-scale studies has been carried out by Dawkins (1976, 1977) concerning hens' preferences for wire-floored cages vs. outdoor runs or litter-covered floor pens. Those experiments illustrate some of the techniques used and difficulties of interpretation in terms of animal welfare.

What do the behavioral tests of Dawkins suggest about hens' preferences for housing? Table 15-1 presents a sample of results obtained. Variable results occurred from one testing situation to another, as previously noted in the thermal and lighting studies with pigs. When kept two hens per cage prior to testing (Experiment 1, 1976), hens entered cages with less delay than when entering the

Table 15-1 *Preference for wire-floored battery cages vs. runs or litter-covered floor pens as influenced by experience and testing condition*

Laying phase, pretest housing	Preference test condition	Response (sec)	
	1976, Experiment 1		
	Time to enter:		
2 hens/cage (12♀♀)	Cage + hens	41	
	Cage	159	
	Run + hens	398	
	Run	1420	
	1976, Experiment 2 (first test)		
	Time to enter:		
2 hens/cage (8♀♀)	Cage	171	
	Run	527	
Outside house and run (8♀♀)	Cage	36	
	Run	6	
	1977, Experiment 1		
	Area preferred, 12 hr		
2 hens/cage (12♀♀)	Cage	5 hens	
	Floor	7 hens	
	1977, Experiment 2		
	Area preferred, 24 choices	First 3 trials	Trials 4–24
2 hens/cage (7♀♀)	Cage	12	35
	Run	9	112
Outside house and run (7♀♀)	Cage	0	9
	Run	21	138

SOURCE: Partial summary from results presented by Dawkins (1976, 1977).

less familiar "run"; hens also entered either type of environment more readily when other hens were present. The importance of previous physical environments was explored more fully in Experiment 2 of 1976; hens that had been kept in outside pens entered either a cage or an outside run more readily than those from cages. However, hens from cages moved more rapidly into cages than into an outside run, and those from runs showed the opposite kind of response.

In her second report, Dawkins found that of 12 hens previously kept in cages, 5 preferred the cage and 7 preferred the floor-pen environment, when allowed a half-day of free access to both (Experiment 1, 1977). In the final experiment considered here (Experiment 2, 1977), hens were kept either in cages or outside before being tested in a simple two-choice situation. During the first three trials, previously caged birds chose cages more often, but outside hens chose the runs exclusively. That difference was significant, but the significance disappeared for the remainder of the 24 trials; even so, previously caged birds chose the runs at a lower rate than those from outside (76% vs. 94%, respectively).

Findings such as those reported above would ordinarily be all that a study would include, but Dawkins (1977) asks: "Do hens suffer in battery cages?" As she then states, many scientists would consider the question to be invalid, as the answer cannot be known directly. How are preference-trial results to be interpreted in trying to answer this question? Dawkins considered a variety of approaches, and Duncan (1977, 1978) and Hughes (1977) have made critical appraisals of those, followed by responses from Dawkins (1979). Behavioral criteria alone appear to allow only subjective judgments of the well-being of animals. Physiological and productivity criteria should also be included in determining the adequacy of an environment. Nevertheless, behavioral measurements may give part of the answer and may provide valuable insights into explaining why symptoms of stress or other adverse responses occur.

When animals do clearly express a preference, the answer given is in relative terms. Several writers have stated or implied that the animal's choice indicates what is "good" and that the choice not taken is "bad." Dawkins (1976) points out this oversimplification; we may be dealing with which of two things the animal finds more desirable (both may have some degree of acceptability) or which is less aversive (neither is agreeable to the animal). The point has also been made that when the animal chooses for itself, it may not choose wisely.

Kilgour (1978a, 1978b) has provided summaries of studies

relevant to animal welfare, particularly those concerned with the humane handling and transportation of livestock. He described a study (Kilgour and Mullord, 1973) in which it was shown that regulations designed to protect calves being transported actually caused unnecessary delay and worked against the welfare of both the animals and the human drivers during the journey. Certainly, gross abuse of animals should be unlawful, but legislation may at times be counterproductive to animal welfare, and if not enforced, may lead to widespread disregard by livestock producers.

REDUCTION OF STRESS: STUDIES WITH HENS

Economic Considerations

Although individual productivity may decline with stress, economic considerations alone do not necessarily discourage or prevent stressful situations. Extremely stressful conditions are likely to cause economic losses, but moderate levels, although reducing well-being and performance of individual animals, may nevertheless provide greater net profit. Grover (1975) prepared a report for the Massachusetts (USA) Society for the Prevention of Cruelty to Animals based on productivity and economic considerations that illustrates this. He assumed that keeping two or three hens per cage as compared to a single hen was likely to increase mortality from about 5 to 12 percent and decrease number of eggs laid from 237 to 222 for a full cycle of laying, yet profits would result only from keeping three hens in each cage under the economic conditions prevailing at that time, as shown in Table 15-2.

Applied animal ethologists face a very real challenge. Animal facilities need better design, and management needs to be improved so that social and other stressors may be reduced without unduly increasing costs of production. Approaches may also involve such techniques as using tranquilizers where severe but short-term stressors are operative, habituating animals or altering their early experience in such ways as to make environments provided by man less aversive, or by selecting genetic stocks that are stress-resistant.

The Changing Environment

The rush of technological innovation in poultry farming has resulted in a transformation of what was essentially a backyard enterprise 50

Table 15-2 *Costs and returns from placing one, two, or three laying hens per cage*

Item	Hens per cage		
	1	2	3
Hens housed	6,660	13,330	20,000
Productivity per hen			
Number of eggs laid	237	222	222
Mortality (%)	5	12	12
Yearly costs and returns per hen			
Fixed costs[a]	$ 3.30	$ 1.66	$ 1.11
Pullet + feed costs	9.40	8.84	8.84
Income	11.08	10.31	10.31
Labor return	−1.62	−0.19	0.36
Labor return, total	−$10,800	−$2,500	$7,200

[a]Fixed costs include the following: building, equipment, electricity, repairs, taxes, and insurance.

SOURCE: Summarized from computations by Grover (1975).

years ago to the largely mechanized production of meat and eggs in many countries today. Because poultry production practices have changed so drastically in recent years, emphasis here will be in seeing the kind of behavioral problems that arise under intensive production systems and what can be done to alleviate some of the problems.

Group Size and Density. Adams and Jackson (1970) cited the results of a number of studies indicating that, in general, egg production of chickens declines and mortality increases as group size increases and area per bird decreases in wire-floored cages. Their own comparisons, based on relatively large numbers of pullets of commercial egg-laying strains in the United States, were relatively consistent with other studies, although having somewhat higher overall mortality. Average results obtained by Adams and Jackson, calculated from two full-year experiments, are shown in Table 15-3. The high mortality experienced in their study involved considerable losses from cannibalism in both years and a disease for which no vaccine was then available (leukosis) in one year.

"Mutilation." The emotion-laden word "mutilation" is sometimes used in describing husbandry practices such as removing a portion of a hen's beak (beak trimming or debeaking) or a chicken's toenails (declawing). Cannibalism among chickens is distressing behavior for both the chicken and the caretaker and can be a serious

Table 15-3 *Group size and density effects on egg production and mortality of hens*

Area per hen (cm^2)	2 or 4 hens per cage	8 or 16 hens per cage
	Rate of egg laying per hen housed (%)	
700	60.8	56.0
350	50.1	44.3
	Mortality over 365-day test period (%)	
700	18.3	23.1
350	21.9	29.9

SOURCE: Calculated from results presented by Adams and Jackson (1970).

problem, particularly under high-density conditions, as was evident in the study of Adams and Jackson (1970). It is a common practice among American poultry producers using multiple-hen cages to remove at least one-half and often two-thirds of hens' beaks to minimize cannibalistic pecking. Failure to do so may result in high mortality from this vice (Chapter 13).

Hens kept in larger groups in cages are likely to be extremely excitable. In the extreme case, they may become "hysterical" (Elmslie et al., 1966; Hansen, 1976). Many poultry farmers who experimented with larger group sizes in multiple-hen cages in earlier years later reduced the number kept in a cage to three or four. Even with as few as three hens in cages and allowing 60 inches^2 (390 cm^2) for each hen, considerable feather loss occurs occasionally, as shown in Figure 15-3. Removal of the middle and inner toenails or "declawing" was found to be of considerable value for hens kept in seven-bird cages by Martin et al. (1976 and personal communication). This practice resulted in 7.8% better livability, 4.2% higher rate of lay, better feed conversion, and improved back feathering.

These examples indicate how removal of certain bodily structures, although causing temporary pain to individuals, can be of much benefit to the welfare of the group.

Housing Changes. Because of the sometimes severe behavioral problems associated with high-density housing, some investigators have begun to experiment with changes in the design of facilities. In the case of multiple-hen cages, "reverse" or "shallow" cages have attracted attention. Reverse cages have feed and water along the long side of the cage and have little depth in contrast to "conventional" cages. Thus, a conventional 12 × 18 inch (30 × 46 cm) cage, holding

Figure 15-3 When hens are kept together in high-density cage environments the amount of feather loss varies considerably from cage to cage. Three 1-year-old hens from the same cage (above) showing considerable feather loss on their backs and wings; they had floor areas of 60 inches^2 (390 cm^2) per hen. Three 1-year-old hens from a different cage (below) with frayed flight and tail feathers, but relatively intact feathers otherwise; they had the same space as those above. Removal of toenails has been shown to reduce feather damage, at least in larger cages holding more hens. (Photographs by the author.)

three or four hens, would have feed and water along the 12-inch front while the reverse cage would have them along the 18-inch front and would be only 12 inches deep. It is reasoned that all birds in a shallow cage should have immediate access to feed and water without having to force their way to the feeder and would be less likely to step on and break eggs that had not previously rolled into the collecting tray. In agreement with this expectation, an advantage for the shallow cage has been found in several studies (Lee and Bolton, 1976; Martin et al., 1976; Swanson and Bell, 1977). However, Hill and Hunt (1978) found opposite results; their hens tended to overeat in shallow cages. It appears that some genetic strains may be better adapted to one cage shape and others to the alternate.

Elmslie et al. (1966) observed hysteria in flocks of 14 hens in wide cages (122 × 46 cm), which was alleviated by subdividing each cage into three units and housing only three hens in each smaller cage. Choudary et al. (1972), using wider and shallower cages (110 × 35 cm) containing 14 hens each, also found that the birds were very excitable and likely to run to one end of the cage and "pile up." If shallow cages are to be used, it appears that they should not be very wide or contain many more than three or four individuals unless genetic stocks that are less excitable are used.

Hysteria in Barren Cage Environments. Apparently, there are different types of nervous and hysterical behavior, which occur at different ages and in different environments. R. S. Hansen of Washington State University carried out a series of experiments (Hansen, 1976) on the type of hysteria that may occur in laying hens, usually after they have been in cages for at least 100 days. This hysterical condition was characterized by "sudden wildly flying about, squawking and trying to hide." Hansen noted that hens in adjacent cages are not necessarily disturbed unless they are also in a hysterical state. Although single birds probably trigger an episode within a cage, the entire group usually reacts almost instantaneously. When hysterical episodes begin to occur regularly within a flock, scratches and torn skin are seen, especially on the back (as in Figure 15–3 or worse); feed consumption and egg production then decrease rapidly.

Hansen found the incidence of hysteria to be closely related to "population pressure"; with flocks of 40, 30, or 20 per cage, he had 91, 50, and 22% hysteria, respectively. Toenail removal reduced hysteria, some genetic stocks were resistant, and certain cage modifications prevented its occurrence.

Changes in Cage Design. Experimental multiple-hen cages have been designed by Hansen (1976), Bareham (1976), and Brantas (shown by Kilgour, 1978b), which differ from conventional commercial cages in providing roosts, nests, and in two of the three, litter or sand baths. As shown in Figure 15-4, Hansen's modified cage is essentially a two-level structure, so that the area available for each hen is essentially doubled when roosting and nest box are included. In his final experiment, Hansen compared two genetic stocks, one of which proved to be hysteria-susceptible and the other much less so. Hen-housed rate of egg production was improved by 4% for the less nervous strain and by 14% for the more nervous strain in cages having roosts and nest boxes.

An experimental cage designed by Bareham (1976) provided an even more natural environment than did Hansen's. It contained litter in nest boxes, which allowed scratching, dust bathing, and more typical nesting behavior. Litter in the nest boxes was moved out

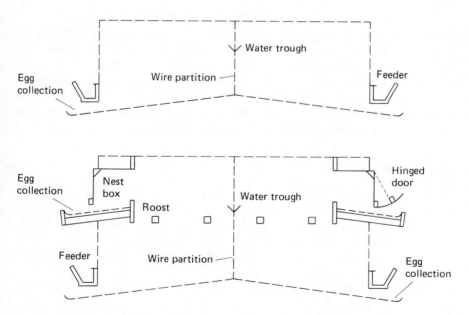

Figure 15-4 Cross-sectional sketch of conventional (above) and modified (below) multiple-hen cages used by Hansen (1976). Hysteria occurred with high incidence when 40 laying hens of certain genetic stocks were kept in conventional cages but not when kept in modified cages at high density. (Sketched from plans supplied by R. S. Hansen.)

frequently, along with the eggs, on a conveyor belt at the time when food was replenished. Litter was recycled after cleaning. Because nearly all eggs were laid in nests, horizontal floors in the nonnesting area were practical. In comparing behavior in conventional barren cages and in modified ones, Bareham found many differences; hens in the experimental cages did much less feather pecking and showed lower frequencies of behaviors believed to be associated with frustration. Greater feather damage, foot problems, and more deaths due to cannibalism were present in conventional cages. Although the number of eggs laid did not differ, hens in experimental cages had heavier eggs. Because the test lasted only 6 months, rather than the 12-month or longer period ordinarily used commercially, differences in productivity due to treatments were probably underestimated. This is likely, as nervousness and hysteria (if occurring at all) are more likely to be evident, with associated extreme losses in productivity in the latter part of the laying year, as found by Hansen (1976).

Whether the kind of cages designed and tested by Hansen and Bareham will be adopted commercially is likely to be determined on the basis of economic considerations in most countries. Although their cages reduced stress and showed promise of increasing productivity, alternative procedures of dealing with behavioral stress, such as debeaking, declawing, or using genetic stocks better suited temperamentally to high-density cage environments, compete as methods for dealing with stressors involved.

Genetic Influences

When different genetic stocks of chickens are intermingled in floor pens, those that are subordinate do less well than if they are kept separately (Tindell and Craig, 1959; Lowry and Abplanalp, 1970, 1972). Relatively nonaggressive strains, as well as subordinate hens within stocks, show evidence of being stressed in terms of performance traits in the presence of more dominant and aggressive birds.

Poultry breeders, and perhaps breeders of other species of livestock, may inadvertently select for increased levels of aggressiveness and social dominance when selecting for early reproductive performance under competitive conditions. Three kinds of evidence that this happens in chickens of egg-laying strains were cited by Craig et al. (1975) as follows: (1) High social status is associated with greater access to feed and other resources and with earlier onset of reproduction; thus socially aggressive and dominant individuals enjoy better environments within the group and being more productive are likely to be selected. (2) Social dominance ability is genetically influenced. (3) Selection for earlier sexual maturity may amount to indirect selection for aggressiveness and dominance as those traits appear to be genetically correlated with age at maturity. Breeders wishing to decrease levels of aggression within their stock may need to select for better temperament (i.e., more peaceful dispositions).

Flightiness or nervousness, and tendencies for hysterical behavior (under stressful social situations) are influenced, at least in part, by inheritance, as shown by Hansen (1976). For example, in comparing two genetic stocks, he found no hysteria in one strain, but it was present in the other stock in three of six 30-hen lots and in five of six 40-hen lots. Al-Rawi et al. (1976) also found evidence that strains differed in their ability to withstand larger-group-size conditions in cages.

HANDLING, MOVEMENT, AND RESTRAINT

How can animals be effectively controlled? In routine husbandry they may need to be moved from one location to another or restrained for purposes such as clipping wool, milking, vaccination, harnessing, saddling, and showing. The ease with which these things can be done varies greatly, depending on inherited differences in temperament, early experience, sex, age, size, strength, social environment, and the like. The expertise of the stockman and adequacy of handling facilities are also likely to be important.

Emphasis in this section will be placed on movement and restraint of animals in situations where they are likely to be ill at ease, frightened, and likely to attempt escape. Handling animals under those conditions has recently been considered in some detail by Kilgour (1978a, 1978b) and Grandin (1977, 1978). Grandin has paid particular attention to how cattle perceive their environment and to psychological influences, such as their desire to remain with the group and to avoid blind alleys, in designing facilities. Her approaches and designs will be examined as models, with the realization that some modifications, other than changes in scale, may be required for other species, depending on their primary senses of perception and characteristic life-styles.

Moving Animals in Familiar Environments

Social status and leader–follower relationships appear to be only moderately associated, if at all, in most species. Although territorial cocks and herd stallions may "drive" their females and young when protecting them from obvious dangers, and females of many species actively lead and protect their young progeny, leadership in other situations is not so clear (Chapter 4). Followership appears to be a fairly strong phenomenon in many species, as most animals resist being left behind. Therefore, if a leading animal can be controlled, often the remainder of a group can be moved easily. The "Judas goat" is well known to many stockmen; it has been used with success in leading flocks of sheep when in strange surroundings, sometimes into slaughterhouses.

The relationship of social position and movement in yearling dairy heifers was studied in three situations by Beilharz and Mylrea (1963). When grazing, those of medium and high status were generally at the front of the group, with actual leaders usually being of medium status. After being released from confinement in an observation yard, the group was likely to be led out by low- or medium-ranking individuals. During forced movements, when driven at a walking pace, very low status heifers were at the front, those of highest dominance tended to be in the middle, and those of medium rank were in the middle or rear. Thus, in general terms, highest-ranking heifers tended to be centrally located within groups during movement. A somewhat similar situation was reported by Kilgour (1969) in a milking parlor; cows of middle and low rank tended to enter first, while those of high rank were found to enter in the middle of the group.

Crowding or forcing gates can be used in encouraging cows to enter the milking parlor. If mechanically driven, gates should be hinged at the top so that they will not injure an animal that has fallen or is unable to move. Kilgour (1978b and personal communication) indicated that crowding gates should not be electrified to shock slow-moving or reluctant animals. When that is done, the gate becomes a "backing gate," and the animals in their anxiety to avoid shocks face it and back away as it approaches.

Movement and Restraint
in Strange Surroundings

In many situations where animals must be handled, there is little or no opportunity for learning. At an abattoir or auction market animals pass through only once, and on large ranches, under extensive management systems, animals are usually moved through yards, chutes, dipping vats, squeezes, or restraining crates only once or a few times during their lifetimes. Under such conditions they become highly aroused and may injure themselves. They also pose problems of possible injury, frustration, and heavy physical exertion by those working with them unless special care is used in handling and adequate facilities are available. Knowledge of animal behavior can be extremely helpful under those circumstances and can transform difficult and potentially dangerous situations into quiet, orderly, and efficient ones (Grandin, 1978).

Sensory Perceptions and Handling. Cattle are easily disturbed by loud or unusual noises. Compressed air is now used in many facilities for operating gates and restraining crates. The hissing noise of escaping air and those sounds made by motors and pumps are likely to cause animals to balk unless they are muffled or kept at some distance. On some commercially available restraining crates, the pump and motor are located on top, an obviously inappropriate place.

Most ungulates have an extremely sensitive sense of smell. Cattle have been observed refusing to enter chutes where others were being castrated or abattoirs when blood was on the floor or when air with the odor of blood was carried to them. Those observations suggest that the odor of blood may be aversive to cattle. If this proves to be true, ventilation systems should be used that direct the odor of blood away from incoming animals.

Cattle have been described as visual animals, and what they see can frighten them. With 360-degree panoramic vision they easily detect movement in any direction. Whether they have any color vision is not clear, but it seems likely that it is only partially present, if at all. What is clearly evident is that cattle often balk and refuse to cross over a shadow or zebra pattern or to walk beneath overhead catwalks. Grandin places special emphasis on solid-sided chutes, which not only block out undesirable strong contrasts of light and shadow in the path to be followed, but eliminate outside distractions as well. A poorly designed long, straight chute is shown in Figure 15–5.

Many stockmen have observed the tendency of cattle to move toward a brightly illuminated area, and Grandin reported that she had seen animals readily enter a lighted abattoir at night where they had to be forced to enter during the day. The abattoir involved solved the balky cattle problem (which existed only during the day) by extending a single-file "race" into the holding pen; cattle moved more readily in single file. Grandin recommends that such races extend at least 15 ft (4.5 m) out from buildings that are to be entered during the day. She also recommends that a light be placed overhead in the stunning pen of abattoirs so that animals will hold their heads up.

Curved Races with Solid Sides. Following recommendations of Rider et al. (1974, in Grandin, 1978), Grandin designed efficient

Figure 15-5 Poorly designed, long, straight chute. Shadows cast by the bars frighten cattle. Chutes with solid sides prevent such strongly contrasting dark and light areas and block out outside distractions. (Courtesy of Temple Grandin and the Brangus Journal.)

cattle-handling facilities that make use of lanes and forcing pens with curved and high, solid sides, as shown in Figure 15-6. Those facilities, although initially costly, are especially useful for slaughterhouses, auctions, and operations where large numbers of animals are handled. Curved races have the advantage that animals will follow those ahead but are prevented from being frightened by seeing people or strange objects until it is too late to turn back. The solid sides also prevent visual distractions from the outside, as noted earlier.

A system incorporating many desirable design features for handling cattle as they are brought in from pasture or range is shown in Figure 15-7. The facility has a goblet-shaped gathering pen which receives the animals. A group of cattle is visualized as a semi-cohesive mass that needs to be made to flow. The rounded shape of the gathering area eliminates corners where cattle would bunch up; as the animals move around the handler in the gathering pen, they will flow along the fence and follow the funnel into the curved holding area or "sorting reservoir." The handler can then block the "neck of the goblet" to prevent animals from returning. The curved holding lane provides access to a "squeeze," a loading ramp, and

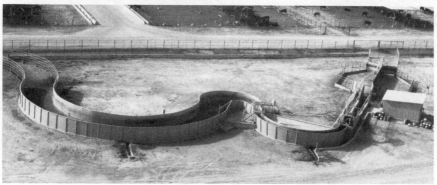

Figure 15-6 Curved, solid-sided races prevent animals from being frightened by seeing what lies ahead and from distractions outside the race. Above: Cattle flow easily into a "sorting reservoir." Below: An efficient facility for handling cattle with a crowding pen (center) leading into a curved single-file chute. Cattle move readily through the curved chutes; handlers work from catwalks. Three people can move 600 cattle per hour through the facility for dipping with a minimum of excitement. (Courtesy of Temple Grandin, Livestock Consultant, Tempe, Arizona.)

diagonal sorting pens. When used for sorting, a single handler either on foot or on horseback can control the flow of animals into the diagonal pens.

Round pens have been used for handling horses, sheep, and wild animals, as well as cattle. Animals that are active and excitable are less likely to injure themselves and others when kept in a circular enclosure. Diagonal sorting pens with gates making wide angles are also useful for avoiding right- or acute-angle corners and blind alleys

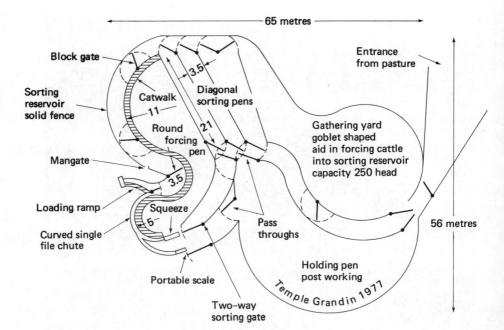

Figure 15-7 Cattle handling facility incorporating many desirable features. (Copyright by Temple Grandin, Livestock Consultant, Tempe, Arizona.)

or dead ends, which can cause extreme avoidance behavior or piling up if other animals coming from behind force others into the area.

Close Physical Restraint

Animals not habituated to close restraint may panic while being caught or held and may injure themselves or the caretaker unless precautions are taken. Familiar sights, sounds, and other animals are reassuring and a quiet approach is desirable. Physical facilities, designed as indicated above, can be very helpful.

Control of the head is especially important in restraining large animals, and simple mechanical devices can be effective. Thus, blindfolds are useful for calming visually oriented horses in some otherwise frightening situations. "Twitches," consisting of strong but lightweight cords or chain loops tightened around the upper lip of a horse, can immobilize the animal. Blunt-ended pincers and nose rings can be used effectively for holding or leading mature cattle, and rope or wire loops ("snares") placed around the upper snout are effective for holding swine (Figure 15-8). Bits and halters are

Figure 15-8 A wire-loop "snare" placed around the upper snout effectively restrains a young boar. (Photograph by the author.)

used with horses and "choke chains" with dogs; pressure is applied for control only as needed.

Whole-body control is possible with smaller animals; foals and young calves may be picked up without injury to them by gathering them in with the arms placed around the chest and behind the hind legs. Puppies and kittens may be safely picked up by the scruff of the neck (as the mother does). Being held tightly in unusual positions often results in a trance-like or "hypnotic" state for many animals, as when sheep are set upon their rumps and held between the legs for shearing and when poultry are held upside down by their legs.

Tonic Immobility or "Animal Hypnosis." Most farm children of the previous generation learned to "hypnotize" chickens. A variety of methods were used, such as tucking the bird's head under its wing and then swinging it back and forth several times before laying it down quietly on its side or back. Typically, when animals are caught and forcefully restrained, three distinct phases of behavior are seen (Gallup, 1974). Initially, they struggle to escape; if that is ineffective they become immobile. Finally, they recover abruptly and again attempt to escape.

Nearly all animals are susceptible to tonic immobilization when forcefully restrained for a few minutes under fear-producing conditions (Ratner, 1967; Gallup, 1974). Darwin was fascinated by this phenomenon and suggested that it served as a defensive response against predation after being caught. The "fear hypothesis" appears more plausible when conditions are considered that increase susceptibility to tonic immobility. Those conditions include: adrenaline injection, naive subjects, isolation from familiar animals, strange surroundings, rough handling, low social status, and the continuing close presence of an apparent predator. Animals that are immobilized have their eyes open and apparently continue to monitor the environment, but they have temporarily lost the power of voluntary movement.

A few experiments have been carried out which tend to confirm the importance of the tonic immobility response as a predator-defense behavior. Thus, Sargeant and Eberhardt (1975) observed under natural conditons that red foxes often lose wild ducks which exhibit tonic immobility after capture and then escape after the foxes lose interest. With captive foxes, they found that 29 of 50 ducks survived initial capture by becoming immobilized and escaping later.

REFERENCES

ADAMS, A. W., and M. E. JACKSON, 1970. Effect of cage size and bird density on performance of six commercial strains of layers. Poult. Sci. 49:1712–1719.

AL-RAWI, B. A., J. V. CRAIG, and A. W. ADAMS, 1976. Agonistic behavior and egg production of caged layers: genetic strain and group-size effects. Poult. Sci. 55:796–807.

BALDWIN, B. A., and D. L. INGRAM, 1967. The effect of heating and cooling the hypothalamus on behaviour thermoregulation of the pig. J. Physiol. 131:375–392.

BALDWIN, B. A., and G. B. MEESE, 1977. Sensory reinforcement and illumination preferences in the domesticated pig. Anim. Behav. 25:497–507.

BALDWIN, B. A., and J. O. YATES, 1977. The effects of hypothalamic temperature variation and intracarotid cooling on behavioural thermoregulation in sheep. J. Physiol. 265:705–720.

BAREHAM, J. R., 1976. A comparison of the behaviour and production of laying hens in experimental and conventional battery cages. Appl. Anim. Ethol. 2:291–303.

BEILHARZ, R. G., and P. J. MYLREA, 1963. Social position and movement orders of dairy heifers. Anim. Behav. 11:529–533.

CHOUDARY, M. R., A. W. ADAMS, and J. V. CRAIG, 1972. Effects of strain, age at flock assembly, and cage arrangement on behavior and productivity in White Leghorn type chickens. Poult. Sci. 51:1943-1950.

CRAIG, J. V., M. L. JAN, C. R. POLLEY, A. L. BHAGWAT, and A. D. DAYTON, 1975. Changes in relative aggressiveness and social dominance associated with selection for early egg production in chickens. Poult. Sci. 54:1647-1658.

DAWKINS, M., 1976. Towards an objective method of assessing welfare in domestic fowl. Appl. Anim. Ethol. 2:245-254.

———, 1977. Do hens suffer in battery cages? Environmental preferences and welfare. Anim. Behav. 25:1034-1046.

———, 1979. Interpreting ethological data. Appl. Anim. Ethol. 5:189-192.

DUNCAN, I. J. H., 1977. Behavioural wisdom lost. Appl. Anim. Ethol. 3:193-194.

———, 1978. The interpretation of preference tests in animal behaviour. Appl. Anim. Ethol. 4:197-200.

EGAN, J. K., J. W. McLAUGHLIN, R. L. THOMPSON, and J. S. McINTYRE, 1972. The importance of shelter in reducing neonatal lamb deaths. Aust. J. Exp. Agric. Anim. Husb. 12:470-472.

ELMSLIE, L. J., R. H. JONES, and D. W. KNIGHT, 1966. A general theory describing the effects of varying flock size and stocking density on the performance of caged layers. Proc., 13th World's Poult. Congr., Kiev, pp. 490-495.

GALLUP, G. G., 1974. Animal hypnosis: factual status of a fictional concept. Psychol. Bull. 81:836-853.

GRANDIN, T., 1977. Practical applications of behavioral principles to the design of cattle handling facilities. Am. Soc. Anim. Sci. Meet. Madison, Wis. (Abstract).

———, 1978. Design of lairage, yard and race systems for handling cattle in abattoirs, auctions, ranches, restraining chutes and dipping vats. Proc., 1st World Congr. Ethol. Appl. Zootech., Madrid, pp. 37-52.

GROVER, R. M., 1975. Poultry Management. A Report to the Massachusetts Society for the Prevention of Cruelty to Animals. Dept. Vet. Anim. Sci., University of Massachusetts, Amherst, Mass.

HANSEN, R. S., 1976. Nervousness and hysteria of mature female chickens. Poult. Sci. 55:531-543.

HILL, A. T., and J. R. HUNT, 1978. Layer cage depth effects on nervousness, feathering, shell breakage, performance, and net egg returns. Poult. Sci. 57:1204-1216.

HOLMES, C. W., and L. E. MOUNT, 1967. Heat loss from groups of growing pigs under various conditions of environmental temperature and air movements. Anim. Prod. 9:435-452.

HUGHES, B. O., 1976. Preference decisions of domestic hens for wire or litter floors. Appl. Anim. Ethol. 2:155-165.

——, 1977. Behavioural wisdom and preference tests. Appl. Anim. Ethol. 3:391-392.

INGRAM, D. L., 1967. Simulation of cutaneous glands in the pig. J. Comp. Pathol. 77:93-98.

——, and K. F. LEGGE, 1970. The thermoregulatory behaviour of young pigs in a natural environment. Physiol. Behav. 5:981-987.

INGRAM, D. L., J. W. HEAL, and K. F. LEGGE, 1975a. Heat loss from young unrestrained pigs in an outdoor environment. Comp. Biochem. Physiol. 50:71-76.

INGRAM, D. L., D. E. WALTERS, and K. F. LEGGE, 1975b. Variations in behavioural thermoregulation in the young pig over 24 hr periods. Physiol. Behav. 14:689-695.

KILGOUR, R., 1969. Social behaviour in the dairy herd. N.Z. J. Agric. 119: 34-37.

——, 1978a. The application of animal behavior and the humane care of farm animals. J. Anim. Sci. 46:1478-1486.

——, 1978b. Minimizing stress on animals during handling. Proc., 1st World Congr. Ethol. Appl. Zootech., Madrid, pp. 303-322.

——, and M. M. MULLORD, 1973. Transport of calves by road. N.Z. Vet. J. 21:7-10.

LEE, D. J. W., and W. BOLTON, 1976. Battery cage shape: the laying performance of medium- and light-body weight strains of hens. Br. Poult. Sci. 17: 321-326.

LOWRY, D. C., and H. ABPLANALP, 1970. Genetic adaptation of White Leghorn hens to life in single cages. Br. Poult. Sci. 11:117-131.

——, 1972. Social dominance difference, given limited access to common food, between hens selected and unselected for increased egg production. Br. Poult. Sci. 13:365-376.

LYNCH, J. J., and G. ALEXANDER, 1976. The effect of gramineous windbreaks on behaviour and lamb mortality amoung shorn and unshorn Merino sheep during lambing. Appl. Anim. Ethol. 2:305-325.

MARTIN, G. A., J. R. WEST, G. W. MORGAN, and T. R. BURLESON, 1976. Effects of wing and toe amputation on layers. Poult. Sci. 55:2061 (Abstract).

MOUNT, L. E., 1968. The Climatic Physiology of the Pig. Edward Arnold (Publishers) Ltd., London.

RATNER, S. C., 1967. Comparative aspects of hypnosis. *In* J. E. Gordon (Ed.), Handbook of Clinical and Experimental Hypnosis. Macmillan Publishing Co., Inc., New York.

ROMAN-PONCE, H. H., W. W. THATCHER, D. E. BUFFINGTON, C. J. WILCOX, and H. H. VAN HORN, 1977. Physiological and production responses of dairy cattle to a shade structure in a subtropical environment. J. Dairy Sci. 60:424–430.

SARGEANT, A. B., and L. E. EBERHARDT, 1975. Death feigning by ducks in response to predation by red foxes (*Vulpes fulva*). Am. Midl. Nat. 94: 108–119.

SWANSON, M. W., and D. D. BELL, 1977. Layer performance in reverse vs. conventional cages. Poult. Sci. 56:1760–1761.

THATCHER, W. W., F. C. GWAZDAUSKAS, J. C. WILCOX, J. TOMS, H. H. HEAD, D. E. BUFFINGTON, and W. B. FREDRIKSSON, 1974. Milking performance and reproductive efficiency of dairy cows in an environmentally controlled structure. J. Dairy Sci. 57:304–307.

TINDELL, D., and J. V. CRAIG, 1959. Effects of social competition on laying house performance in the chicken. Poult. Sci. 38:95–105.

WATSON, R. H., G. ALEXANDER, I. H. CUMMING, J. W. MacDONALD, J. W. McLAUGHLIN, D. J. RIZZOLI, and D. WILLIAMS, 1968. Reduction of perinatal loss of lambs in winter in western Victoria by lambing in sheltered individual pens. Proc. Aust. Soc. Anim. Prod. 7:243–249.

16

Sexual Behavior: Brain and Hormone Effects

Stereotyped views of what is "normal" sexual behavior are common, but close observation of what actually occurs reveals that males may exhibit what are viewed as female behaviors, and vice versa. Thus, Geist (1971) found in bachelor bands of sheep in a natural habitat that subordinate males would stand to be mounted by dominant rams. Such behavior is not uncommon among young bulls and boars kept in all-male groups. A large accumulation of unpublished data by University of Nebraska researchers Jakway and Sumption (cited by Wells, 1966) indicated that boars reared together formed stable homosexual pair relationships which persisted even after they were placed with sows and both males were copulating repeatedly with females. The passive member of those pairs would stand quietly, allowing rectal intromission and ejaculation. The occurrence of homosexual activity between young boars was confirmed by Wells, who found it to be especially frequent in a group kept in close confinement.

The frequency of mounting and thrusting activity by females varies greatly among species; it is seen infrequently among hens, with moderate frequency in dogs, and regularly among cows. A survey by Beach (1968) revealed that such male-like sexual behavior of adult females had been reported in 13 species of mammals distributed among five taxonomic orders. Indeed, mounting of cows by other cows is used by many cattle producers who do not keep bulls as the most reliable index for indicating which cows are in heat and are therefore approaching ovulation and should be artificially inseminated (Figure 16-1).

Figure 16-1 One cow attempts to mount another, but the animal underneath moves away, indicating that she is not in estrus. Cows in heat may attempt to mount others but are more commonly those willing to stand while others mount. The sexually aggressive animal (the one mounting) need not be in heat herself. (Photograph by the author.)

That hormones produced by the testes and ovaries are involved in mating behavior is clearly evident, as removal of the gonads typically reduces or eliminates those behaviors. However, animals castrated after a certain sensitive period can be returned to mating activity typical of their sex by injecting androgens in males and estrogens or estrogens plus progesterone in females. Such replacement techniques are useful in determining the physiological requirements of animals for particular hormones in maintaining sexual activity.

MALES

Castration Studies

Castrated males, although lacking sperm, may continue for some time to mount, have penile erection, achieve intromission, and show ejaculatory reflexes when stimulated by receptive females. Variation in the retention of mating behavior is large, especially among ungulates and carnivorous mammals, and some individuals may continue to show elements of mating behavior for more than 1 to 3 years. A survey of the literature by Clemens and Christensen (1975) indicated that all males of domestic fowl stopped copulating within 5 weeks of castration and rodents and rabbits within 5 to 13 weeks. Some castrated dogs and cats continued to mount receptive females for at least 3 years. However, there was a reduction in the frequency of intromissions and amount of time spent in the "locked" position of Beagle males after 6 months (Beach, 1970). Experimental studies of ungulates are meager, but bulls have been reported to continue mounting for up to a year after castration (Folman and Yolcani, 1966).

It has been suggested that the maintenance of sexual behavior following castration may be due to secretion of androgen from the adrenal cortex or because of previous sexual experience. However, surgical removal of the adrenals does not appear to influence postcastrational behavior in either dogs (Schwartz and Beach, 1954) or cats (Cooper and Aronson, 1958), and lack of experience did not prevent postcastrational sexual performance of male Beagles (Hart, 1968). However, Rosenblatt and Aronson (1958) found that male cats with sexual experience continued to mate for a longer period than inexperienced ones; some experienced tom cats were still able to achieve intromission 10 weeks after castration, but none of those lacking experience succeeded after the first week. A breed difference in persistence of mounting and intromission following castration of male rabbits was found by Agmo (1976). He also found that although inexperienced males were about equal in mating behavior to experienced males for the first 2 or 3 weeks after castration, their performance dropped off much more rapidly thereafter.

When castrated males that have stopped copulating are injected with testosterone, there is a latency period before complete mating behavior is recovered. In most species studied the recovery period is 1 or 2 weeks, but pigs may require 6 weeks (Joshi and Raeside, 1973).

A fascinating development of the 1970s was the establishment of the concept that testosterone may be a "prehormone" which does not act directly, but is converted to the appropriate metabolite when it reaches target tissues. Testosterone is converted to estradiol, at least in some species, in the central nervous system (Naftolin et al., 1975; Selmanoff et al., 1977). Surprisingly, estradiol, previously thought of as a "female" sex hormone, is effective in inducing some components of male sexual behavior in laboratory animals (see Kelley and Pfaff, 1978). It has also been learned that estradiol acts synergistically with another metabolite of testosterone (dihydrotestosterone) in restoring copulatory behavior in the castrated male rat (Baum and Vreeburg, 1973).

Masculinization of the Brain

Some mammalian species, such as rats and golden hamsters, are still at an early stage of development at birth. Neonatal castration of such early-stage males has profound effects on their behavioral responses to hormones later in life. Thus, male rats castrated soon after birth release gonad-stimulating hormones from the pituitary gland cyclically as females do, and they exhibit female sexual postures when injected with estrus-inducing hormones (reviewed by Gorski, 1979). Such castrates also fail to show male sexual behavior when injected with androgens as adults. However, if castration is delayed for only a few days after birth, such individuals do respond as males when androgen-stimulated later. Therefore, it may be concluded that there is a short period, very early in life, during which the genetic male must be exposed to a minimal amount of androgen to become a functional male later.

Presumably, the male's own testes secrete sufficient androgen to cause masculinization early in life. Early castration has similar effects in several species, including the dog, in which the young are born relatively undeveloped. However, the effects are less pronounced in dogs, probably because they are further along in development at the time of birth (Beach, 1975). Early castration of the precocial species of domestic animals does not have the same kind of effect. Males of precocial species have already been masculinized prior to birth.

Where are the sites of masculinization, and what hormone is involved? Various kinds of evidence indicate that the hypothalamus and preoptic areas of the brain are important (Morrell and Pfaff, 1978; Gorski, 1979), but precise brain locations vary somewhat among species. At least in the rat, testosterone appears to be converted to an estrogenic hormone within the central nervous system, and it is estrogen that causes the brain to become organized so that male-like behavior will be exhibited at later ages. (It will be seen later that masculinization of the female brain also occurs if it is supplied with either testosterone or estrogen during the critical period.)

What about birds? Because the embryo is in an egg outside the body of the mother, it is readily accessible to hormonal treatment either by being dipped in hormone solutions or by being injected. Wilson and Glick (1970) used both routes of administration and then attempted to activate precocial sexual behavior by injecting androgen into juvenile chicks. Both waltzing and mating behavior of treated males were suppressed as compared to that of males untreated during embryonic life. Feminization of embryonic males of Japanese quail also results if they are similarly exposed (Adkins, 1975); crowing, strutting, mounting, and cloacal contact were all greatly suppressed. Therefore, results are opposite in avian species to those found in mammals; exposure of the embryonic chick to either testosterone or estradiol feminizes genetic males.

General Hormonal Effects

Androgens secreted by the interstitial cells of the testes are largely responsible for the development of secondary sexual traits of males during adolescence and for activation of mating behavior beginning at puberty and continuing thereafter. Immediate control of sex hormone production resides in the gonadotropic hormones produced by the anterior pituitary gland. Pituitary secretion of gonadotropins is controlled in turn by releasing factors from the hypothalamic region. Figure 16-2 illustrates some of the major interrelationships between hormonal and environmental influences in males. The output of follicle stimulating hormone (FSH) and luteinizing hormone (LH) is relatively constant in the male as contrasted to the female. The names of the gonadotropic hormones derive from their functions in the female (where they were first studied in depth). FSH has the primary function of supporting production of sperm cells in the testes. LH causes the interstitial cells to produce male sex hormones.

Androgens cause the production of pheromones by glands and tissues. The pheromones are odorous substances that may be rubbed as glandular excretions onto vegetation or conspicuous objects, may be carried by the urine, or evaporate directly into the air. In any case they are readily detected by other animals of the same species. They are useful for identifying the animal's presence (or previous presence) or physiological state. They have particular importance in territorial marking by males and may indicate reproductive readiness in either sex. Androgens make the male's glans penis epithelium more sensitive to tactile stimulation, thereby increasing the promptness of ejaculation when the penis contacts the female's copulatory organs (Aronson and Cooper, 1968).

Levels of male sex drive in guinea pigs appear to be largely influenced by differences in reactivity of individuals to androgen stimulation. Grunt and Young (1953) showed that not only did individual male guinea pigs differ in sexual activity when intact, but those differences appeared again following castration and androgen therapy. Although a minimal amount of androgen was required for their castrated males to return to the precastrational level of mating, dosages above that did not increase activity further. Even with very high lev-

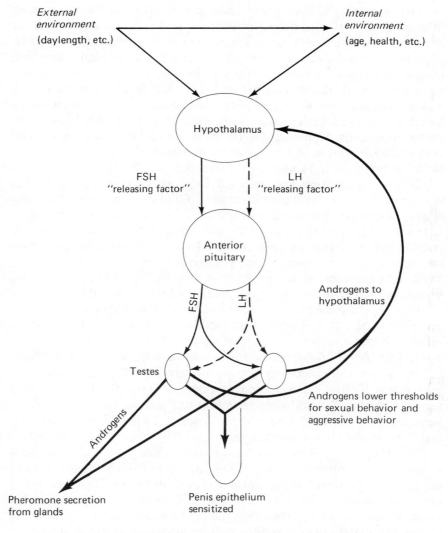

Figure 16-2 Some hormonal and environmental influences on reproductive traits and behavior in the male.

els of androgen stimulation, males found to be low in sex drive before castration continued to be low. These and other results obtained by Young's group with guinea pigs support the concept that males with low libido have nervous mechanisms and tissues associated with sexual arousal and mating behavior that are refractory to androgen stimulation.

Other studies indicate that androgen insufficiency may, at least in some genetic stocks, be responsible for low male sex drive and that androgen treatment can cause some previously nonbreeding males to copulate and produce progeny as effectively as naturally breeding males. As a graduate student at the University of Wisconsin I worked with a strain of albino rats having a high incidence of nonbreeding males. Bioassays involving seminal vesicle and prostate weights of breeder and nonbreeder males were used to indicate levels of circulating androgens. Comparisons of those glandular weights suggested that relatively small, but real, differences existed in androgen output between breeders and nonbreeders (Craig et al., 1954).

When testosterone was injected daily at different levels, mating with ejaculation was induced in about 40% of previously nonbreeding males, although untreated nonbreeders continued their celibacy. Latency until first ejaculation ranged from 5 to 15 days, which is comparable to latencies encountered by others with castrated male rats receiving replacement therapy. Dosage levels appeared to be relatively unimportant, as the percentage responding at very high levels was no greater than that of males receiving lesser amounts.

Our results (Craig et al., 1954) indicated that frequencies of completed matings, pregnancy rates, and the number of young sired by induced-breeder males were at least equal to that of naturally breeding males (see Table 16-1). Thus, there was substantial evidence of simple androgen deficiency in the previously nonbreeding males that responded. Those males failing to respond (60% of those treated) were refractory to androgen stimulation, just as the low-sex-drive guinea pigs studied by Young's group were. Either kind of situation may be responsible for low sex drive.

Whether stimulation with exogenous androgen or with gonadotropins in attempting to elicit breeding of low-libido males in farm animals can be justified needs careful consideration. If genetic influences are responsible, the answer should surely be negative under most circumstances. On the other hand, if psychological or previously traumatic experiences are responsible, such an attempt might be worthwhile. It will be shown below that social situations can have profound effects on levels of circulating androgens, either in causing increases under favorable circumstances or reductions for individuals under stress.

Short-Term Changes in Androgen Levels

Circumstantial evidence indicated for a long time that androgenic hormone levels of males are not cyclic in the same manner as in fe-

Table 16-1 *Frequencies of successful matings (with ejaculation), pregnancies, litters with living young, and litter size for natural and induced-breeder males*

| Male breeder classification | Number of males | Treatment period | | Treatment and post-treatment periods | | | |
		Total mating trials	Females bred (%)	Total females bred	Pregnant (%)	Litters of living young (%)	Litter size
Natural	10	150	37.2	58	89	68	6.4
Induced	11	111	52.9	113	87	72	6.5

SOURCE: Calculated from results presented by Craig et al. (1954). Data from inbred and noninbred matings have been combined.

males. Although long-term changes such as those associated with the seasons are evident for some species, shorter cycles of several days to several weeks duration (as in polyestrous females) are obviously lacking. With the advent of sensitive biochemical assay methods, it has now become apparent that males do show short-term changes in LH and testosterone in the peripheral circulation, but those rhythms are relatively rapid, with several peaks and troughs occurring daily, as in rabbits (Moor and Younglai, 1975) and bulls (Katongole et al., 1971).

The importance of sexual stimulation in affecting short-term changes in LH and testosterone levels is being explored. Whether rapid changes in LH levels are caused by sexual stimulation is controversial, but significant increases in testosterone levels have been found following brief exposures of intact males to sexually receptive females in mice, rats, hamsters, rabbits, bulls, and Rhesus monkeys (Hutchinson, 1978). Exposure to vaginal secretions of estrous females caused about the same increase in circulating testosterone as did copulation in hamsters (Macrides et al., 1974). The functional significance of these rapid increases in androgen levels (and perhaps in LH levels) is not yet clear.

Longer-Term Changes in Androgen Levels

Seasonal changes, or artificial changes mimicking those, especially changes in the proportions of light and darkness per day, are associated with changes in testicular activity in some mammalian and many avian species living outside the tropics. For example, when male quail are changed from short day lengths to long ones, they show increases in LH and testicular growth (Follett and Farner, 1966).

The effects of social environment on androgen secretion have been measured directly and also indirectly by examining testicular and male accessory gland growth. Vandenbergh (1971) used the indirect method with young male mice that were kept from the time of weaning in groups having no adult present, an adult male, or an adult female. Because mice, as well as many other mammals, are sensitive to olfactory stimuli, those mice kept with adult females were housed in a separate room from the others, to prevent exposure of young males of the other groups to adult female pheromones. Males were then removed at intervals and necessary measurements made. Testicular development was most rapid when males were reared in the presence of an adult female, but the presence of an adult male had an inhibitory effect on testicular and accessory gland development. Vandenbergh's casual observations on behavior suggested that the adult male inhibitory effect was not due to fighting or abuse of the young males; there were no physical signs of injury. Young males kept in groups without an adult male (but in the same room as adult males) were intermediate in rate of sexual development. The immediate cause of the mature animals' stimulatory and inhibitory effects on young males' androgen levels (as reflected by testes and gland weights) is not clear, but these phenomena could be of practical importance in domesticated species.

Testosterone levels in the blood have been measured for adult male Rhesus monkeys kept in different social contexts. Higher levels were found for more dominant or aggressive individuals, but it was not clear whether those levels were present before social status was determined or were the result of it (Rose et al., 1971). In a second study, Rose et al., (1972) attempted to determine whether testosterone levels of individual males were relatively stable or not by moving them from one situation to another. When placed one at a time with a group of females in a large outdoor compound, males easily established social dominance and engaged in frequent sexual activity. During 2 weeks of intimate contact with the females, each of four adult males had increased levels of plasma testosterone with two- to three-fold increases over that of the preceding and following 2-week periods when they were individually caged. The same males were subsequently placed, one at a time, with a socially organized group of 30 adult males. Within minutes they were attacked, and attacks continued until the introduced males were rescued 2 hours later. Plasma drawn 3 to 5 days after the brief, but decisive, defeat was significantly lower in testosterone. Tests run 6 weeks later on two individuals showed a continuing depressed level of testosterone. That situation was reversed when the same two males were again placed, one at a time, with the female group; by the fourth day circulating testosterone had returned to the same high level found during the previous heterosexual grouping. Because the male monkeys, while with the females, were both dominant and sexually active, the relative importance of those two influences on androgen levels could not be separated. What is clear is that testosterone secretion can be either increased or decreased significantly by social variables, and those changes may persist for periods of at least several weeks. A possible argument to account for such changes in group-living animals is that they may be adaptive. Rose et al. (1972) stated: "As aggressive threats or challenges by a subordinate animal are severely punished by the dominant members of a group, high levels of testosterone stimulating such behavior could be viewed as inappropriate and maladaptive. In a parallel vein, increase in testosterone stimulated by access to females would function to support the increased frequency of sexual activity."

FEMALES

Behavioral Differentiation

The very young mammal whose brain is unstimulated by androgen or estrogen during the sensitive period for masculinization has what may be described as a "female" brain, in terms of behavioral responsiveness, regardless of its sex chromosomes. Just the opposite situation appears to occur in avian species; the embryonic bird has a "male" brain in behavioral potential unless feminized by androgen or estrogen during the critical period.

Freemartins. Cattle producers know that most heifer calves born as twins with bulls will be sterile but that the male member of the pair will be fertile. Such abnormal calves, although having external organs of female type, have internal organs of reproduction that are predominantly male-like in character (Lillie, 1917). Romans of the first century B.C. referred to such animals as *taurae* or "female-bulls," but they are now known as freemartins. Until early in the present century it was not known whether these abnormal animals were genetic males or females. Absence of sexual receptivity as expected of normal adult females added to the confusion. Tandler and Keller in Europe and Lillie in the United States, working independently, discovered that anatomical abnormalities typical of sterile freemartins occurred when blood vessels connected the circulatory systems of an embryonic bull calf and the abnormal animal (Marcum, 1974). Lillie found three typical female embryos in 24 cases where a male was also present in the uterus, but in each of those three the blood systems of the embryos were *not* joined. On the basis of his findings, Lillie (1917) proposed that:

1. Sterile freemartins are genetic females that shared their embryonic blood circulation with a male.

2. The embryonic male's testes become active in secreting androgen before the female's ovaries become functional.

3. Masculinization of the heifer calf occurs as a result of embryonic exposure to androgen.

Data combined from numerous studies indicate that about 92%

of all heifers born as a twin with a bull calf will be sterile (Marcum, 1974). The sterile freemartin also occurs in heterosexual twins of sheep, but the incidence is only about 1%. The situation may be encountered in goats, pigs, and horses also, but the event appears to be rare.

Masculinization. Studies by Harris and others indicated that it is the central nervous system of the very young female mammal that is affected by male hormone rather than the pituitary or the ovaries. When pituitary glands or ovaries from "masculinized" females were transplanted to nonmasculinized females, they functioned in female fashion (Harris, 1964).

Studies with females of several mammalian species support the hypothesis that testosterone transformed (aromatized) to estrogen in the central nervous system during the early sensitive period causes masculinization (Goldman, 1978), as it also does in the genetic male (see earlier). Removal of the ovary during this early period does not change the potential of the genetic female to secrete gonadotropins cyclically or to show typical female behavior in later life when injected with female sex hormones. Thus, the young mammal (either female or male) whose central nervous system receives less than a certain amount of androgen or estrogen during the sensitive period will have brain tissues organized in female fashion.

Hormones and Mammalian Estrous Cycles

Figure 16-3 indicates some major effects of the female sex hormones and how pituitary hormones and hypothalamic releasing factors control their production. The hypothalamus has a major role in controlling the "estrous cycle." For seasonally breeding species, day length is of major importance in initiating or terminating reproductive activity. Developmental and social factors (such as the presence or absence of a male) can also modify the onset or duration of sexual receptivity. Copulation is required in some species before ovulation occurs, but that is not common in domesticated animals.

Under the influence of signals received from the hypothalamus, the pituitary gland secretes FSH and LH in cyclic fashion. FSH causes one or more ovarian follicles, containing maturing eggs, to develop. The enlarging follicle itself becomes an endocrine gland of major importance; it secretes the female sex hormone or estrogen. Estrogen has multiple effects, one of which is to act in a feedback mechanism on the hypothalamus so that at an appropriate time the pituitary gland is stimulated to release a surge of luteinizing hormone or LH,

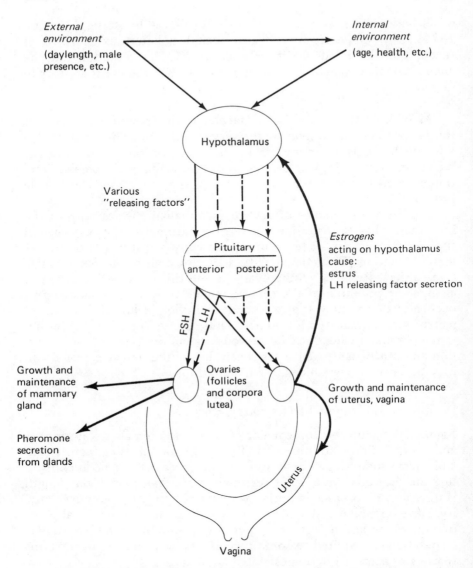

Figure 16-3 Some hormonal and environmental influences on reproductive traits and behavior in the female.

which causes the mature follicle(s) to rupture, thereby releasing the egg(s) into the reproductive system. After ovulation the ruptured follicle develops a yellow body or "corpus luteum," which, similar to the developing follicle that preceded it, becomes an endocrine gland and secretes the hormone progesterone.

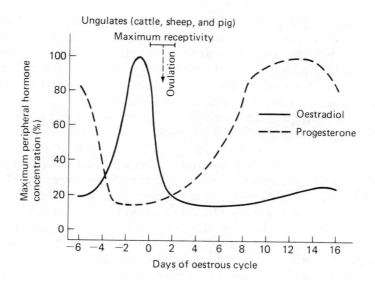

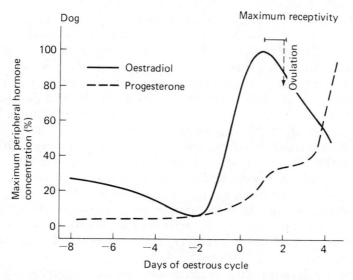

Figure 16-4 Percentage of maximum estrogen (estradiol) and progesterone in the peripheral blood plasma during the estrous cycles of ungulates (above) and the dog (below). Day 0 is the day when maximum receptivity occurs. (From Clemens and Christensen, 1975.)

Female sexual receptivity (estrus or heat) is closely integrated with ovulation, so that a male, if present, is readily accepted and sperm are delivered into the reproductive tract at an optimal time for fertilization. Figure 16-4 shows how the period of maximum female receptivity and ovulation are associated with estrogen (estradiol) and progesterone secretion for certain ungulates and the dog. The typical pattern of hormonal secretion differs somewhat among species, but the period of maximum receptivity and ovulation follows peak levels of estrogen secretion in all species by half a day or more.

If fertilization and implantation of embryos do not occur, poly-estrous females proceed through repeated cycles of ripening follicles, with attendant rises in circulating estrogen, sexual receptivity, ovulation, and anestrous periods. On the other hand, if pregnancy is established, the placenta that forms about and nourishes the embryos produces progesterone, as the corpora lutea also continue to do. Continuing high levels of progesterone inhibit renewed FSH secretion (via brain centers), and the chain of events necessary for continuing cycles and periodic estrous periods is broken.

The role of progesterone in the onset and termination of estrus varies considerably among species and appears to be of major importance among seasonal breeders, such as sheep, in priming ewes to exhibit estrus during the breeding season. Lisk (1978) has extensively reviewed what is known about the hormonal regulation of estrus, with particular emphasis on the effects of progesterone. He makes it clear that although a high level of estrogen must be present before estrus is expressed, progesterone can either facilitate or inhibit its occurrence. Timing relationships are important; for example, simultaneous injection of estrogen and progesterone does not induce heat in any species studied. Among ungulates such as the cow, sow, and ewe, onset of heat does not occur until 1 to 3 days after progesterone has reached the low point of its cycle (Figure 16-4). Among induced ovulators such as rabbits and cats (see below), ovulation does not occur, and corpora lutea are not formed during the breeding season unless mating occurs. In these animals progesterone levels remain low even though estrogen levels and sexual receptivity show cyclic changes as ovarian follicles mature, and then in the absence of copulation, regress.

First ovulations after puberty are not accompanied by sexual receptivity in 100% of ewes and 75% of heifers (Lisk, 1978), and the first ovulation of the breeding season is also "silent" in terms of estrus expression in the ewe (Thorburn et al., 1969). Thus, only *after* progesterone has been secreted by a first set of corpora lutea will the ewe show heat under the influence of estrogen. Lisk asks whether this mechanism is peculiar to sheep or whether this species only provides an extreme example of a general function of progesterone. He cites a study (Carrick and Shelton, 1969) involving heifers tested by the presence of bulls for expression of sexual heat, which may be interpreted in the same way (i.e., that progesterone primes the female for normal responsiveness to estrogen in the expression of heat).

Copulation and Ovulation

Among domesticated mammals, the cat and rabbit are "reflex" or "induced" ovulators; in contrast to most other species, they normally ovulate only in response to mating. If mating does not take place, the mature ovarian follicles and the enclosed eggs regress. Rabbits in estrus sometimes ovulate in response to mounting by another female, but Diakow (1971) has shown that intromission and genital stimulation of the sexually receptive female is necessary to cause ovulation in the cat. That a nerve pathway from the genital tract is directly involved was demonstrated by Diakow; none of five females ovulated in which vaginal and uterine nerves were severed, but four of five sham-operated controls did, following genital stimulation during heat.

Although male presence and copulation are not required for ovulation in laboratory rodents or ungulates, evidence is accumulating which indicates that such stimuli may increase promptness of ovulation, proportion of eggs fertilized (if mating is done artificially), likelihood of successful implantation, and maintenance of pregnancy. Obviously, such effects are not of major importance in cattle, where artificial insemination is used, possibly because of the considerable genital stimulation that results from the technique. However, efficient detection of sexual receptivity in the absence of a bull is a serious problem in some herds. Sows and ewes are clearly dependent on stimuli from sexually active males for early and full expression of heat and for maximum reproductive performance. Consideration will now be given to some male influences that are associated with hormonal changes or where that is clearly implied; other social effects on reproductive behavior will be dealt with in Chapter 18.

Copulation and Hormone Release

Ovulation from mature follicles is dependent on a surge of LH from the pituitary gland in response to a releasing factor from the hypothalamus. The pituitary response may be due to either an increase in releasing hormone from the hypothalamus or because the pituitary becomes more sensitive to a constant level of the releasing hormone (reviewed by Smith and Davidson, 1974). Regardless of which mechanism is operative, copulation, at least in the rat, can stimulate LH release over and above that which would occur spontaneously, and the level of response is related to the intensity of the sexual behavior (Moss and Cooper, 1973). Copulation was found to delay the spontaneous decline in plasma LH that would otherwise occur, and Rodgers (1971) found that an increased number of ova were shed when females mated at that time. He suggested that the additional ova came from mature follicles which would not have ruptured until a later cycle in the absence of copulation and the continued LH surge associated with it.

Boars that engage in more courting activity, especially nosing the flanks of sows in heat before mating, have higher conception rates (Hemsworth et al., 1978). In view of the findings in rats, it is reasonable to ask whether LH levels may have been involved. However, Hemsworth et al., although not excluding that possibility, suggested that extra flank nosing may have stimulated oxytocin release from the sow's pituitary gland; such a hormonal release was postulated as increasing sperm transport, number of sperm in the oviduct, and chances of fertilization. Estrous goats have higher oxytocin levels when a male is present, as shown by McNeilly and Drucker (1972).

We appear to be on the threshold of new understanding of relationships among social stimuli, hormonal changes, and reproductive success. Although lacking hormonal assay data, a study on the importance of vaginal stimulation in relation to pregnancy in the hamster by Diamond (1972) illustrates this. Diamond artificially inseminated a number of estrous females by placing sperm directly into the uterine horns (bypassing the vaginal route). Of 24 females receiving vaginal stimulation prior to artificial insemination, 80% delivered living young. However, of 20 females inseminated in the same way, but without vaginal stimulation, only 15% produced litters with living young. Fertilization via the uterine horns occurs whether the vagina is stimulated or not. Diamond postulated that natural mating causes neural impulses to be transmitted from the vagina to the hypothalamus with several consequences: normal estrous cycling is interrupted, continuing progesterone production is supported (indirectly), and a "biological clock" is set in motion which eventually leads to successful parturition. Although research is needed in domestic animals to check on the generality of his results, what evidence is available appears to support Diamond's statement that "no longer can sexual behavior and copulation be looked at simply as a means of bringing sperm to egg; rather, the behavior must be considered an indispensable and crucial component to be considered in the reproductive process and in the evaluation of reproductive failure."

Components of Estrous Behavior

A case has been made by Beach (1976) for separating the complex events of estrus into three separate categories of attractivity, proceptivity, and receptivity. He emphasized the need to recognize that sexual behavior of the female is intimately associated with stimuli and activity provided by the male partner; that sexual attraction is mutual, both sexes being likely to exhibit appetitive behavior; and that the consummatory phase also involves both sexes (Chapter 5). Those behaviors are profoundly affected by the hormonal environment, and it is readily apparent when sexually mature males and females are kept together that the female becomes attractive to the male and shows appetitive behavior (proceptivity) before she will stand for sexual intercourse. Whether the subdivision of estrous behavior into the three arbitrary classifications suggested by Beach will be widely adopted is not yet clear, although it has been used in considering sexual behavior as related to hormonal influences in female primates (Herbert, 1978). As with any classification system, there are often difficulties in deciding how a particular kind of activity should be categorized. For example, Beach (1976), in discussing attractivity (defined in terms of male responsiveness to a female), indicates that proceptivity (sexually appetitive behavior of the female) increases the male's responsiveness; females that actively "solicit" copulation have a higher stimulus value. Beach points out that a female's appetitive behavior is a reflection of the male's attractivity. Proceptive behavior, in Beach's system, is also likely to be exhibited even as the female "receives" the male. Thus, sows and females of other species push backward during copulation, thereby assuring maintenance of intromission, and receptive bitches also show proceptivity and receptivity at the same time by moving their vulvas so that insertion can be achieved if the male dog's pelvic thrusts are misdirected so that intromission does not occur at once.

One method of learning about the effects of hormones on attractivity as opposed to appetite for sexual intercourse is to restrain one animal while allowing another freedom to approach or not, or to choose the company of another animal under different hormonal influence. This approach was used by Le Boeuf (1967), who learned that tethered bitches were visited by males on 93% of days when they were in estrus as compared to 75% on days before and after estrus. In the reverse situation, male dogs were visited on 76% of the days by estrous females, but on only 35% of days when in the nonestrous stage. A study of estrous behavior of ewes by Lindsay and Fletcher (1972) used the same general approach. They found that about 75% of intact ewes visited a tethered ram during a particular estrous period, but all ewes eventually visited a tethered ram some-time during the mating season. When ovaries were removed and dos-ages of estrogen were increased well above levels required to induce estrus, proceptivity was increased so that all of 60 ewes visited a tethered ram.

The attractivity of boars and associated proceptivity of sows has been studied by Signoret and his associates using a T-maze technique (Signoret, 1970). Thus, females approached a boar more frequently when in estrus, but were just as likely to approach animals of either sex at other times.

REFERENCES

ADKINS, E. K., 1975. Hormonal basis of sexual differentiation in the Japanese quail. J. Comp. Physiol. Psychol. 89:61-71.

AGMO, A., 1976. Sexual behaviour following castration in experienced and inex-perienced male rabbits. Z. Tierpsychol. 40:390-395.

ARONSON, L. A., and M. L. COOPER, 1968. Desensitization of the glans penis and sexual behavior in cats. *In* M. Diamond (Ed.), Perspectives in Reproduc-tion and Sexual Behavior. Indiana University Press, Bloomington, Ind.

BAUM, M. J., and J. T. M. VREEBURG, 1973. Copulation in castrated male rats following combined treatment with estradiol and dihydrotestosterone. Sci-ence 182:273-285.

BEACH, F. A., 1968. Factors involved in the control of mounting behavior by female mammals. *In* M. Diamond (Ed.), Perspectives in Reproduction and Sexual Behavior. Indiana University Press, Bloomington, Ind.

——, 1970. Coital behavior in dogs. VI. Long term effects of castration upon mating in the male. J. Comp. Physiol. Psychol. Monogr. 1-32.

——, 1975. Hormonal modification of sexually dimorphic behavior. Psychoneu-roendocrinology 1:3-23.

——, 1976. Sexual attractivity, proceptivity, and receptivity in female mammals. Horm. Behav. 7:105-138.

CARRICK, M. J., and J. N. SHELTON, 1969. Oestrogen–progesterone relation-ship in the induction of estrus in spayed heifers. J. Endocrinol. 45:99-109.

CLEMENS, L. G., and L. W. CHRISTENSEN, 1975. Sexual behaviour. *In* E. S. E. Hafez (Ed.), The Behaviour of Domestic Animals, 3rd ed. Baillière Tindall Publishers, London.

COOPER, M., and L. ARONSON, 1958. The effect of adrenalectomy on the sex-ual behavior of castrated male cats. Anat. Rec. 131:544 (Abstract).

CRAIG, J. V., L. E. CASIDA, and A. B. CHAPMAN, 1954. Male infertility associated with lack of libido in the rat. Am. Nat. 88:365-372.

DIAKOW, C., 1971. Effects of genital desensitization on mating behavior and ovulation in the female cat. Physiol. Behav. 7:47-54.

DIAMOND, M., 1972. Vaginal stimulation and progesterone in relation to pregnancy and parturition. Biol. Reprod. 6:281-287.

FOLLETT, B. K., and D. S. FARNER, 1966. The effect of the daily photoperiod on gonadal growth, neuro-hypophysial hormone content and neuro-secretion in the hypothalamo-hypophysial system of the Japanese quail. Gen. Comp. Endocrinol. 7:111-124.

FOLMAN, Y., and R. YOLCANI, 1966. Copulatory behaviour of the prepubertally castrated bull. Anim. Behav. 14:572-573.

GEIST, V., 1971. Mountain Sheep: A Study in Behavior and Evolution. University of Chicago Press, Chicago.

GOLDMAN, B. D., 1978. Developmental influences of hormones on neuroendocrine mechanisms of sexual behaviour: comparisons with other sexually dimorphic behaviours. In J. B. Hutchison (Ed.), Biological Determinants of Sexual Behaviour. John Wiley & Sons, Inc., New York.

GORSKI, R. A., 1979. The neuroendocrinology of reproduction: an overview. Biol. Reprod. 20:111-127.

GRUNT, J. A., and W. C. YOUNG, 1953. Consistency of sexual behavior patterns in individual male guinea pigs following castration and androgen therapy. J. Comp. Physiol. Psychol. 46:138-144.

HARRIS, G. W., 1964. Sex hormones, brain development and brain function. Endocrinology 75:627-648.

HART, B. L., 1968. Role of prior experience in the effects of castration on sexual behavior of male dogs. J. Comp. Physiol. Psychol. 66:719-725.

HEMSWORTH, P. H., R. G. BEILHARZ, and W. J. BROWN, 1978. The importance of the courting behaviour of the boar on the success of natural and artificial matings. Appl. Anim. Ethol. 4:341-347.

HERBERT, J., 1978. Neuro-hormonal integration of sexual behaviour in female primates. In J. B. Hutchison (Ed.), Biological Determinants of Sexual Behaviour. John Wiley & Sons, Inc., New York.

HUTCHISON, J. B., 1978. Hypothalamic regulation of male sexual responsiveness to androgen. In J. B. Hutchison (Ed.), Biological Determinants of Sexual Behaviour. John Wiley & Sons, Inc., New York.

JOSHI, H. S., and J. I. RAESIDE, 1973. Synergistic effects of testosterone and oestrogens on accessory sex glands and sexual behaviour of the boar. J. Reprod. Fertil. 33:411-423.

KATONGOLE, C. B., F. NAFTOLIN, and R. V. SHORT, 1971. Relationship between blood levels of luteinizing hormone and testosterone in bulls, and the effects of sexual stimulation. J. Endocrinol. 50:457–466.

KELLY, D. B., and D. W. PFAFF, 1978. Generalizations from comparative studies on neuroanatomical and endocrine mechanisms of sexual behaviour. *In* J. B. Hutchison (Ed.), Biological Determinants of Sexual Behaviour. John Wiley & Sons, Inc., New York.

LeBOEUF, B. J., 1967. Interindividual associations in dogs. Behaviour 29:268–295.

LILLIE, F. R., 1917. The free-martin; a study of the action of sex hormones in the fetal life of cattle. J. Exp. Zool. 23:371–423.

LINDSAY, D. R., and I. C. FLETCHER, 1972. Ram-seeking activity associated with oestrous behaviour in ewes. Anim. Behav. 20:452–456.

LISK, R. D., 1978. The regulation of "heat." *In* J. B. Hutchison (Ed.), Biological Determinants of Sexual Behaviour. John Wiley & Sons, Inc., New York.

MACRIDES, F., A. BARTKE, F. FERNANDEZ, and W. D. ANGELO, 1974. Effects of exposure to vaginal odor and receptive females on plasma testosterone in the male hamster. Neuroendocrinology 15:355–364.

McNEILLY, A. S., and H. A. DRUCKER, 1972. Blood levels of oxytocin in the female goat during coitus and in response to stimuli associated with mating. J. Endocrinol. 54:399–406.

MARCUM, J. B., 1974. The freemartin syndrome. Anim. Br. Abstr. 42:227–239.

MORRELL, J. I., and D. W. PFAFF, 1978. A neuroendocrine approach to brain function: localization of sex steroid concentrating cells in vertebrate brains. Am. Zool. 18:447–460.

MOOR, B. C., and E. V. YOUNGLAI, 1975. Variations in peripheral levels of LH and testosterone in adult male rabbits. J. Reprod. Fertil. 42:259–266.

MOSS, R. L., and K. J. COOPER, 1973. Temporal relationship of spontaneous and coitus-induced release of luteinizing hormone in the normal cyclic rat. Endocrinology 92:1748–1753.

NAFTOLIN, F., K. J. RYAN, I. J. DAVIES, V. V. REDDY, F. FLORES, Z. PETRO, and M. KUHN, 1975. The formation of estrogens by central neuroendocrine tissue. Rec. Prog. Horm. Res. 31:295–319.

RODGERS, C. H., 1971. Influence of copulation on ovulation in the cycling rat. Endocrinology 88:433–436.

ROSE, R. M., T. P. GORDON, and I. S. BERNSTEIN, 1972. Plasma testosterone levels in the male Rhesus: influences of sexual and social stimuli. Science 178:643–645.

ROSE, R. M., J. W. HOLADAY, and I. S. BERNSTEIN, 1971. Plasma testosterone, dominance rank and aggressive behaviour in male Rhesus monkeys. Nature 231 : 366–368.

ROSENBLATT, J. S., and L. R. ARONSON, 1958. The decline of sexual behavior in male cats after castration with special reference to the role of prior sexual experience. Behaviour 12 : 285–338.

SCHWARTZ, M., and F. A. BEACH, 1954. Effects of adrenalectomy upon mating behavior in castrated male dogs. Am. Psychol. 9 : 467–468 (Abstract).

SELMANOFF, M. K., L. D. BRODKIN, R. I. WEINER, and P. K. SUTERI, 1977. Aromatization and 5 alpha-reduction of androgens in discrete hypothalamic and limbic regions of the male and female rat. Endocrinology 101: 841–848.

SIGNORET, J. P., 1970. Reproductive behaviour of pigs. J. Reprod. Fertil., Suppl. 11 : 105–117.

SMITH, E. R., and J. M. DAVIDSON, 1974. Luteinizing hormone releasing factor in rats exposed to constant light: effects of mating. Neuroendocrinology 14 : 129–138.

THORBURN, G. D., J. M. BASSETT, and I. D. SMITH, 1969. Progesterone concentration in the peripheral plasma of sheep during the oestrous cycle. J. Endocrinol. 45 : 459–469.

VANDENBERGH, J. G., 1971. The influence of the social environment on sexual maturation in male mice. J. Reprod. Fertil. 24 : 383–390.

WELLS, B. H., 1966. The effect of group vs. individual confinement on the reproductive and behavioral patterns in young boars in pasture-lots and on concrete. M.S. thesis, Virginia Polytechnic Institute Library, Blacksburg, Va.

WILSON, J. A., and B. GLICK, 1970. Ontogeny of mating behavior in the chicken. Am. J. Physiol. 218 : 951–955.

17

Sexual Behavior of Males

This chapter focuses briefly on stimuli releasing male mating behavior and on the promptness, duration, and frequency of mating of males of some major species of domestic animals. However, more emphasis is placed on effects of early social environments, pretesting for serving capacity in natural matings, social status effects, and the "buller-steer" syndrome. Much of the information has been gathered recently, evidence is quite limited in some studies, and new questions arise as others are answered. Because of those qualifications, a number of studies are described briefly (this is also done in Chapters 18 and 19). This approach should give greater insight into the results and the strengths and weaknesses of the evidence underlying tentative conclusions can then be judged.

EARLY EXPERIENCE

Imprinting and sexual attachment of young birds to a different species can have major effects on mating behavior after sexual maturity; mammals are probably affected to a lesser degree (Chapter 9).

Few studies on social isolation from an early age have been carried out with domestic animals, but those which have indicate that this treatment can impair mating efficiency after sexual maturity. There may be a sensitive period, prior to sexual maturity, during which proper orientation, positioning or movement during mounting is learned. Some previously isolated males have difficulty in achieving intromission when mating opportunities arise, even though their sexual excitement is obviously high. With the adoption of early separation of the sexes and placing of animals in individual compartments during the rearing period, under certain new management systems, there is a need to be alert to the possibility that future reproductive capacity may be reduced. A few relevant studies will now be considered.

Chickens

Brown Leghorn cockerels were reared in physical and visual isolation and compared with others reared in an all-male flock from 8 weeks until both treatment groups were fully developed at 6½ months (Wood-Gush, 1957). Normal mating behavior was shown by 2 of 4 isolated males, and by 6 of 10 group-reared ones. During brief testing periods, considerable aggressive behavior toward females was observed for cockerels of both treatment groups, and it may be that failure of some males to achieve clear-cut dominance may have interfered with their sexual behavior. The attainment of successful copulation was not very high in either group, for whatever reason. On the basis of this small-scale study it appears that cockerels can copulate successfully, although reared in solitary confinement from the midjuvenile period until confronted with females when sexually mature.

Cockerels were separated from pullets at 8, 10, or 12 weeks of age, with subsequent confinement in individual cages for 4 to 5 months by Siegel and Siegel (1964). Those males were then compared for mating behavior with others that had been kept in all-male flocks from the same ages. Mating frequency was clearly impaired for those separated at 8 or 10 weeks, but the effect of isolation was diminished considerably when delayed until cockerels were 12 weeks old.

It would appear desirable in future studies such as these to include comparisons with males reared in heterosexual groups and to extend the isolation treatment to earlier ages.

Mammals

Dogs. Beach (1968) compared the mating behavior of male Beagles reared in three environments. One group, designated as "semi-isolated," was weaned at 3 weeks and then kept in individual cages. Another group was weaned at 6 weeks and similarly kept in individual cages of the same design and in the same room as the semi-isolated dogs. Each dog was allowed access to an exercise area for 15 minutes daily; semi-isolated dogs exercised by themselves, but animals of the second group were released along with others so that they could interact physically for the same length period. A third group, consisting of both males and females, was raised from about 6 or 7 weeks of age in a large outdoor pen.

Dogs reared in cages were tested for mating behavior with sexually receptive females; males kept outdoors were tested with females of their own group which were separately penned from proestrus until no longer receptive. Beach concluded that male dogs reared in the semi-isolated environment were less efficient copulators than those of the group reared in individual cages but experiencing 15 minutes of social contact daily. It may be worthwhile to note that effects of age at weaning are possibly confused with the effects of post-weaning social experience (or absence of social experience) for these two groups. Males of the second group, kept in small individual cages but permitted brief daily social contacts, mated as efficiently as those reared in the heterosexual group in the large outdoor pen.

Semi-isolated males achieved intromission in only 24% of mating tests, but males of the second and third groups had success rates of 58 and 54%, respectively. The poor performance of the semi-isolated dogs was clearly *not* caused by lack of interest, as they mounted at about the same frequency as males of the other groups. Of five semi-isolated animals, two mated effectively from the first set of trials, one deteriorated in performance, and the other two failed to show improvement with experience. Two of the three deficient males tended to mount from the rear, but one clasped the female so far back that his genital region never came into contact with the female's; the other male nearly always failed because of lateral stepping movements. The third deficient male consistently mounted abnormally from the side; even when dislodged by the bitch, which would then position herself directly in front, he would persist in moving to the lateral position.

Swine. A study involving rearing environments of boar pigs from 3 to 30 weeks of age by Hemsworth et al. (1977) resembled Beach's in several ways, but important differences in results were obtained. Hemsworth's group was motivated to do their study because of reported low levels of sexual activity of boars in intensively managed piggeries in Australia. In many of those, young boars to be used as breeders are housed in individual pens to allow growth and feed-use evaluation. The boar pigs were reared in three ways: (1) individually, in social restriction with no visual or physical contact; (2) in a group of boars only; and (3) in a group that also contained immature females. Boars in the first two treatments could hear and smell immature gilts.

From 30 weeks of age onward boars were kept individually in pens that were close to mature gilts. At regular intervals each male was evaluated in a mating test and those were continued over a 6-month period. The results, in terms of total successful copulations (with ejaculations) for each treatment group, are shown in Figure 17-1. Socially restricted boars were seriously impaired in their mating behavior as compared to those reared in all-male or male-female groups. The significant reduction in copulation frequency of restricted boars was accompanied by reduced frequency of mounting and shorter average duration of ejaculation.

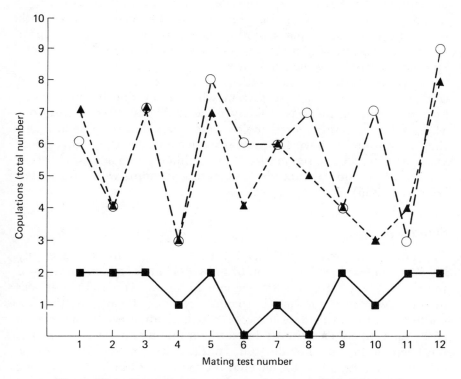

Figure 17-1 Total number of copulations achieved by each group of boars at each of the 12 mating tests. Groups shown as ■——■, group 1 (social restriction); ○----○; group 2 (all-male group); ▲·····▲, group 3 (mixed sex group). (From Hemsworth et al., 1977.)

In comparing their results with other studies on laboratory species and the dog, Hemsworth et al. noted that their boars were visually as well as physically isolated, but physical isolation only appears to have been used in the other studies. Whether that difference can account for the very different results in terms of frequency of mounting and improper clasping of females must await further study; the socially restricted boars mounted only about 30% as often as group reared ones, but when mounting was achieved, it was with normal orientation (i.e., from the rear). Mating efficiency did not improve over time, as is evident from Figure 17-1.

Other Mammals. Although experimental data are not cited, Schein and Hale (1965) reported that young bulls reared without the opportunity to mount other animals lack the "perceptual motor components" needed for appropriate spatial orientation. Thus, they mount a cow from any direction, but are likely to improve with experience. All other parts of the male copulatory pattern appear to be normal, but with intromission prevented by improper position, ejaculation into the vagina cannot occur.

Harlow's famous study with isolation-reared monkeys yielded similar findings to other studies in which mounting frequency was high but fertilization was unlikely because of improper orientation. Thus, when isolation-reared males were compared with feral males they were found to have adequate sex drive and thrusting movements were similar, but insemination was unlikely; females were often grasped by the side or shoulder and mating attempts were disoriented (Harlow, 1965).

Summary

The effects of early experience on male mating behavior have only begun to be explored. Isolation of individual males and separation of the sexes at very early ages are becoming ever more common practices under newer selection and husbandry systems. Results of the few studies conducted to date indicate that individual isolation from an early age is likely to seriously impair later mating efficiency of males, but keeping them in all-male groups has relatively minor or no adverse effects. If males are to be selected on the basis of individual feeding tests and then used for natural matings when mature, they will be more satisfactory as breeders if kept in groups except during feeding.

It now appears that individual isolation interferes primarily with sex drive in cocks and boars, but faulty orientation is the primary defect in mating behavior resulting from lack of social experience in dogs, bulls, and male Rhesus monkeys.

RELEASERS

Turkey toms show courtship displays during the breeding season to models of a hen's body in the crouching position, and a model hen's head mounted on a stick at the proper level above the floor is adequate for releasing mounting and treading behavior (Schein and Hale, 1965). That such behavior is not learned is suggested by similar behavior of isolation-reared young male poults after being primed by androgen injections. In ungulates an adequate sign stimulus for mounting behavior is a dorsal surface with supports (Fraser, 1968: 27-30).

The efficient collection of sperm for use in artificial insemination requires some knowledge of releasers for mounting and ejaculation. Although most males will respond with rather minimal stimulation, better responses can be elicited by skillful handling to minimize anxiety and to maximize arousal and the sensual reward associated with a successful ejaculation (Foote, 1974). Hand mating (temporarily placing together of a breeding male and an estrous female) requires much the same skills and knowledge for good results.

A satisfactory stimulus object is important in eliciting prompt and full expression of a male's breeding behavior. A female in standing estrus would appear to be the ideal subject, but restrained nonestrous females are capable of bringing forth vigorous mating behavior in bulls (Figure 17-2). Restrained steers are often used in bull studs, and "dummies" are also commonly used in the artificial collection of semen from bulls, rams, stallions, and boars. Dummies often bear only a superficial resemblance to an estrous female of the species and yet are mounted willingly by most males. Artificial vaginas, either attached to the dummy, or more commonly hand held, complete

Figure 17-2   A restrained heifer serves as a releaser for a bull being tested for "serving capacity" even though she is sexually nonreceptive. (Courtesy of G. H. Kiracofe, Kansas State University.)

287

the necessary stimulus complex required for ejaculation. Artificial vaginas must be used with attention to proper lubrication, temperature, internal liner pressure, and position for optimal results (Foote, 1974). A temperature of 45°C, well above vaginal temperatures of estrous cows and mares, apparently serves as a supernormal releaser for bulls and stallions. Ejaculations into an artificial vagina may sometimes be obtained at that elevated temperature from males that would otherwise fail to complete the mating act.

Although most males respond to "artificial" females for collection of semen, some sluggish breeders may require estrous females, at least during early stages of semen collection, to release the desired behavior. Courting behavior, including contact (nosing, sniffing, pushing, etc.), strutting or prancing, auditory signals (such as the boar's "courting chant"), and similar behaviors may be associated with adequate arousal. Although these acts may seem to be of little consequence to the human handler, they may be important in preparing the male for subsequent phases of mating. More experienced males tend to abbreviate their courting behavior if they are consistently used in stud matings with females that are already sexually receptive when first encountered or in artificial collection situations where there is no requirement for eliciting an immobile stance of the object to be mounted.

SPECIES-SPECIFIC MATING BEHAVIOR

Males of the various domesticated species vary considerably in the time required to complete a mating and in the number of services that may be completed before they become totally refractory to the attractions of sexually receptive females (Alexander et al., 1974). It is clear, however, that most will usually mate again if a different receptive female is introduced. Males being used in artificial insemination centers are commonly collected from only a few times weekly, although they would willingly copulate more frequently; additional collections rapidly deplete sperm reserves, so that the law of diminishing returns operates.

Less information is available on male mating behavior in other major species of domestic animals than for chickens, sheep, and cattle. However, Wilson et al. (1977) found hybrid vigor for frequency of mating behavior in boars. Nearly 80% of 7½- to 9-month-old crossbred boars mated every time they were exposed to estrous gilts, as compared to about 30% of comparable purebred boars under the same test conditions.

It is difficult to arrive at general recommendations as to appropriate sex ratios for breeding the various species of domestic animals. Genetic differences in sexual behavior, use or nonuse of short-term mating tests, age and degree of sexual development, social-status effects (in multiple-male groupings), amount of space and topography of the area used by the herd or flock, and other variables may all affect the number of females that a male can effectively serve during a breeding period. Even so, some commonly used sex ratios for the major food- and fiber-producing animals are presented in Table 17-1.

Duration of intromission and rapidity of ejaculation vary greatly among domestic animals. Cloacal contact is brief in gallinaceous birds and ejaculation occurs quickly, lasting only a few seconds. Bulls, rams, and male goats achieve intromission rapidly, although several mounting attempts may precede it and ejaculation occurs with a "pelvic thrust" usually within a second following intromission. Horses require about 40 seconds to complete ejaculation and before withdrawal of the penis. Boars may maintain intromission for as long as 20 minutes, although 5 minutes appears to be more typical. Copulation in the dog involves a special physiological phenomenon characteristic of the Canidae (wolves, coyotes, foxes, etc.) in which a portion of the penis swells rapidly after intromission and is held by the vaginal muscles so that a "lock" or "tie" is formed. After a minute or so the male commonly dismounts and faces in the opposite direction, although the two animals remain tied together for 10 to 20 minutes; ejaculation continues until near the end of the period (Fox and Bekoff, 1975). As indicated earlier, cats are of special interest among domestic animals (along with rabbits) in that intromission and ejaculation, although lasting only 5 to 10 seconds,

Table 17-1 *Sex ratios commonly used in major species of farm animals*

Species	Sex ratio (females per male)	Comments
Chicken	14–16	White Leghorns
	10–12	Heavy breeds
Sheep	20	Ram lambs
	25–30	Yearling rams
	33–50	Mature rams
Cattle	25–33	Extensive and harsh conditions
	50–60	Pretested bulls
Swine	12	Less than 1 year old
	20	Older boars

trigger ovulation by the female (Fox, 1975). However, repeated matings within a short period commonly occur in cats. It has been suggested that courtship, fighting, and displays of the tom cat preceding copulation may promote follicle maturation in the ovary, thereby preparing the female.

FERTILIZING CAPACITY
AND PRETESTING

Sexually mature males vary considerably in their sex drive and mating dexterity. Although apparently free of anatomical abnormalities and disease and possessing sperm capable of fertilizing ova, a significant number of males are nevertheless partially or completely infertile because of defective sexual behavior. Genetic variation for mating frequency of males, when tested in the absence of social dominance effects, has been demonstrated by selection responses in chickens (Siegel, 1965; Wood-Gush, 1960) and its heritability is estimated as being high in beef cattle (Blockey, cited by Beilharz, 1978).

There is a serious infertility problem when natural matings are used in "improved" strains of turkeys. Those strains were changed by selection to have extremely broad breasts and fast growth. Turkey toms with those characteristics apparently have reduced mating dexterity and perhaps reduced libido. Studies by Smyth and Leighton (1953) and Hale (1955) indicate that infertility in turkeys is largely a function of both female receptivity (she determines how often matings will occur) and of male dexterity. Clumsy toms terminate hens' receptivity if eversion of the oviduct occurs even though intromission and ejaculation do not occur (Hale, 1955). Hens having experienced an orgasm associated with an incomplete mating become as refractory to further mating as if they had been inseminated. Because of lengthy nonreceptive periods following incomplete matings in many hens, fertility declines in matings where clumsy toms are involved. Increasing the number of males present in a turkey flock having an infertility problem caused by awkward males would be less beneficial than replacing particular males causing the problem.

Broiler breeds of chickens are commonly mated with 8 or 10 cocks per 100 hens while Leghorn-type cocks (egg-laying strains), being lightweight, agile, and high in libido, are commonly used at ratios of 6 or 7 per 100 females. Thus, it appears likely that selection for rapid gains and meat-type conformation may have been counterproductive with respect to reproductive capacity of the male. Beilharz (1978) has suggested that a similar situation may prevail in farm mammals where selection is directed at productivity traits other than those involving reproductive capacity.

Fewer breeding males could be used if they could be rapidly and accurately characterized as to potential sexual effectiveness by some type of screening procedure before joining with the females. It should be noted in this regard that males with extremely high libido may have their sperm reserves depleted rapidly, as found in cockerels (Parker et al., 1940) and in bulls (Almquist and Hale, 1956). Nevertheless, males with higher frequencies of mating in test situations generally produce higher fertility when exposed to large numbers of sexually receptive females.

Chickens

It would be useful if sexual potential of males could be predicted by performance at early ages. Wood-Gush (1963) attempted such an evaluation by injecting immature male chicks with testosterone propionate and testing their mating behavior with a freshly killed chick in a crouching position. The same individuals were tested without hormonal stimulation as adults under standard conditions. Mating behavior in those two situations was uncorrelated.

Methods of testing for sex drive vary considerably in effectiveness. Justice et al. (1962), in reviewing the literature and from their own results, recommended for testing cocks that:

1. Adolescent males should not be placed with females much older than themselves because failure to achieve passive dominance over the females can seriously interfere with sexual behavior.

2. Males should be allowed a period of mating experience prior to the testing interval.

3. A period of 2 or 3 days of sexual deprivation immediately preceding testing periods allows a significant buildup of sex drive and should probably be included in testing.

4. Males should be kept adjacent to or in a cage within the test pen prior to testing, because acquaintance with the test environment (including females?) increases mating frequency.

Using techniques comparable to those recommended, White Leghorn cockerels were tested in terms of scores based on four brief mating trials of each male (McDaniel and Craig, 1959). Several measures of semen quality were also obtained. Fertility tests were then carried out using both multiple-sire and individual-sire matings. Groups of males selected for uniformity of scores ranging from very low to very high values were placed in mass matings in flocks of hens, with results as shown in Table 17-2. Although the fertility levels were relatively low under the conditions of the test (male/female ratio of 1:26 and eggs being collected for fertility testing soon after males were added), such a situation constituted a critical test of the hypothesis that "sexual effectiveness" scores should reflect actual mating values. Whether there may be an optimal score, somewhat less than that of the highest-scoring males, would require further testing (the next-to-the-highest-scoring group had the highest fertility).

In a second test of the value of sexual effectiveness scores, cockerels were placed in single-sire matings, and eggs were collected for fertility testing soon after. Lower-scoring males had a slower buildup of fertility as compared with those having high and medium scores. It was also learned that an index based on both behavioral and semen characteristics would have indicated potential fertilizing capacity better than either measurement alone.

When cocks were observed for frequency of matings in flocks where they were constantly present, it was observed that matings were four times more frequent in late afternoon as compared to early morning hours (Craig and Bhagwat, 1974). Using those results as a basis of calculation, Craig et al. (1977) estimated that half-year-old White Leghorn males were averaging not less than five completed matings per day. Similar estimates based on year-old males indicated a mean frequency of at least 10 matings per day (Kratzer and Craig, 1980).

Table 17-2 *Percentage fertility obtained from males in mass matings when selected on the basis of their sexual effectiveness scores*

Pen	Mean sexual effectiveness score for five males[a]	Percentage fertility[b]
B	0.2	32
C	4.4	39
E	7.6	53
D	10.6	71
A	15.4	63

[a] A high numerical score indicates a high level of libido in the test situation.
[b] Fertility of eggs on days 7 and 8 after groups of males were placed in mating pens.
SOURCE: After McDaniel and Craig (1959).

Sheep

Many rams of the Romney Marsh and Southdown breeds are capable of impregnating over 100 ewes in an 18-day period, as was shown by Haughey (cited by Mattner et al., 1967). When each of 40 rams was placed singly with groups of 200 to 300 ewes under New Zealand conditions, the average number lambing from an 18-day mating period was 160. Nevertheless, with other breeds and under more severe conditions, 100 ewes per ram is obviously too many. Lightfoot and Smith (1968) found that in the south Western Australia "wheat belt," single-ram matings with 25, 40, or 50 ewes gave higher proportions of ewes mated during the first 14 days and also over a total period of 6 weeks. In one comparison they found that ratios of 1:100, 1:50, and 1:25 resulted in 64, 75, and 91% of ewes giving birth after a 6-week mating period.

Multiple-ram matings are commonly used under extensive husbandry conditions and provide some insurance against infertility of individual males caused by low libido or other problems. Fowler and Jenkins (1976) vasectomized some rams and placed both fertile and infertile rams in experimental flocks varying in size from 66 to 133 ewes (sex ratios were always 1 ram:33 ewes). Those flocks were compared with "control" groups having 3 fertile rams and 100 ewes. After a 34-day period (ewes should have been in estrus twice if not conceiving during the first cycle) all ewes were examined for pregnancy. During the mating period most ewes had been marked, presumably mounted, and possibly inseminated by more than one ram; 40, 75, and 95% had been so marked in flocks with 2, 3, and 4 rams, respectively. When infertile rams were socially dominant, pregnancy was reduced by 18% as compared to control groups where all males were fertile. When fertile rams were dominant, pregnancy rates did not differ from the controls, although one-third to two-thirds of subordinate rams were infertile. Fowler and Jenkins pointed out that more adverse effects would probably have been found under more extensive conditions where more ewes would probably have been mated by only one ram.

Besides sex ratios and ram infertility, other factors may also affect fertility and males' breeding activity; relative uniformity of rams' ages gives subordinate rams more opportunity to mate than when rams of mixed ages run together. Very young rams may be unable to mate at the same rate as older ones. Thus, Lightfoot and Smith (1968), although finding no differences in fertility when 1½-year-old rams were compared to 3½- to 5½-year-old rams in flocks with sex ratios of 1:25, were clearly unable to match the older groups' breeding activity when the sex ratio was 1:50.

Do rams distribute their sexual favors equally among ewes, or do they show preferential mating? The answer may depend in part on whether ewes compete for the male's services. If rams tend to ignore particular estrous ewes, for whatever reason, while mating others repeatedly, there would be additional reason for multiple-male matings. However, if rams are strongly attracted to "fresh stimulus objects" in their mating activity, it would be feasible to use single-ram matings more extensively, especially if rams could be rapidly pretested for level of libido and effectiveness of mating as well as for fertilizing capacity of their semen. It has been shown (Pepelko and Clegg, 1964) that sex drive in rams that have already served a number of times is increased when estrous ewes not previously bred are encountered.

Cattle

Most bulls have a high sex drive and can serve several cows daily. Three bulls placed individually with heifers that had been hormonally treated to synchronize estrus were observed continuously by Pexton and Chenoweth (1977). Each of those completed 20 or more matings during the first 24-hour period after being placed with the heifers. Repeated trials revealed that over 90% of all females exhibited estrus within a 30-hour period and that pregnancy rates achieved in groups of 10 to 25 synchronized heifers were comparable to those obtained by two bulls kept in single-sire matings with 25 and 26 untreated heifers. The bulls were distributing their matings among heifers; otherwise, relatively few would have become pregnant, especially when 25 heifers were available per bull.

Although individual bulls usually mate at high frequency when the opportunity exists, some with high libido are inefficient in completing their matings. Rupp et al. (1977) compared bulls in single-sire matings with multiple-sire matings (two or four bulls per group); sex ratios were 1:44 and 1:60 in the single-sire matings. Five of seven bulls in individual-sire matings achieved fertility rates comparable to those in matings with two or four bulls per group of heifers. However, two of the seven in individual-sire matings had very poor conception rates because of poor dexterity; one bull mounted 173 times but successfully copulated on only 6% of his mounts, as compared to 36% conceptions after mounting for the other bulls. Therefore, mounting frequency alone is an inadequate criterion of a bull's mating ability.

That bulls vary greatly in mating ability has been shown by Blockey (1978). He compared the results of short-term mating tests with pregnancy rates achieved in subsequent field trials of bulls in single-sire matings. Of 20 bulls, four with low test scores (0, 1, or 2) established pregnancy in 3 to 67% of their cows, as compared to conception rates of 89 to 100% for those with higher scores. Further field testing revealed that of over 600 bulls given the short-term "yard test," 13% had low scores.

Although schemes for short-term testing of libido and mating capacity in bulls were devised as early as 1959 by Hultnäs and have been under development since (reviewed by Osborne et al., 1971, and Blockey, 1976), it appears that really effective testing procedures have only recently been developed by Blockey (1978). A series of testing procedures were tried and the method described below appears to be both effective and practical.

1. Nonpregnant, nonestrous females are placed in restraining crates (permitting access by bulls) around the perimeter of a pen.
2. A group of bulls is allowed to watch other bulls mounting the restrained heifers for a brief period.
3. The group of bulls is admitted into the yard with the restrained heifers and the number of completed matings for each bull is recorded for a 40-minute test period.

Blockey tested 30 herds in the fashion indicated above. The number of matings completed (i.e., serving capacity) varied from 0 to 19 for Hereford and from 0 to 16 for Angus bulls; wide ranges of scores were obtained in all herds. Bulls were also examined for leg and penis problems. As expected, bulls with deviated penes, healed hematomas of the penis, arthritis, and overgrown claws had high incidences of individuals with low scores. Such bulls might well be eliminated prior to further testing. Nevertheless, of 338 bulls apparently free of such defects, 11% were classified as being in the low-serving-capacity class (with scores of 0, 1, and 2).

General Considerations

From the material presented above, it is seen that the effectiveness of male mating behavior has been significantly altered in some stocks, such as in broad-breasted strains of turkeys, by extreme genetic selection for traits that cause deviations from the body form favored in natural environments. In those stocks and where natural matings are undesirable for other reasons, such as the danger, expense, and lesser genetic gains associated with keeping bulls on individual dairy farms, artificial insemination may replace natural matings. A male's mating dexterity and libido become of secondary importance in situations where artificial insemination is used, if he is an outstanding animal otherwise. Recent changes in husbandry and in insemination techniques may favor wider use of artificial insemination in other livestock systems in the future, as seems likely with swine and broiler chickens.

In systems of production where natural matings are required or desirable, studies indicate that most males have sufficient sex drive and mating dexterity to fertilize more females than are commonly allotted to them. Excessive males are frequently used in multiple-male matings because some are ineffective breeders. Effective methods of pretesting for natural mating ability have been developed for cocks and bulls. Although proven effective, such tests have not been adopted with chickens, apparently because of the expense of testing and the relatively small economic value of individual cocks. The situation is very different where individual males are more valuable and it is likely that testing, such as is currently gaining favor for bulls of beef breeds, may be extended to other species. Thus, as the study of Rupp et al. (1977) made clear, bull/cow ratios of 1:60 involving single-sire matings gave as good results as obtained in multiple-sire matings with ratios of 1:25 with the exception of two of seven individual-bull matings. Observations of mating behavior of those two indicated difficulty in achieving completed services. It is likely that such bulls would have been eliminated by using serving capacity tests, as described above.

When multiple-male matings are required, as in extensive range operations, it may be necessary to keep more males, partly because of social-status effects. Even so, early elimination of males with low mating ability and keeping fewer but better ones is desirable, as costs would be less and genetic gains greater. Also, interferences between males during matings would probably be reduced, as each would be required to serve a larger number of females.

SOCIAL STATUS EFFECTS

Effects of social status on reproductive success in natural populations have already been described in general terms (Chapter 4). But more quantitative information is needed for domesticated animals. In particular, it is important to know to what extent agonistic behavior and interference between males during mating activity reduce fertility.

Chickens

Two small-scale studies carried out with chickens more than 30 years ago are considered as classics (Guhl et al., 1945; Guhl and Warren, 1946) and prompted similar experiments in other species. Results of those studies will be examined and compared with results of more recent ones.

Guhl et al. (1945) carried out extensive observations on mating behavior in flocks of White Leghorns. Among other findings, it was shown that the social status of each of four cocks when kept in an all-male group was not related to mating frequency when tested in flocks of hens individually; thus, the second-ranking male completed more matings than the first-, third-, and fourth-ranking males, and those three did not differ significantly from each other. The situation was very different when all four males were placed together with a group of seven hens. In that situation, agonistic interactions among the males were so severe that the hens remained on roosts most of the time for the first 11 days. The top-ranking male showed intense antagonism toward the second-ranking male and because of his aggressive behavior and interference with mating of his subordinates was described as having "psychologically castrated" them.

Two additional flocks were studied by Guhl and Warren (1946). Males to be placed together in flocks were selected on the basis of producing sperm that competed successfully in producing chicks when hens were artificially inseminated with mixed semen. Information was thereafter obtained by actual observations of agonistic and mating behavior and by the number of chicks sired by each male as indicated by genetic markers (plumage color or comb type). The use of genetic markers was facilitated by using males of different breeds, and breed differences in social dominance may have had large effects. In both flocks the top-ranking male was most successful in observed frequency of mating and in siring the most chicks, while the lowest ranking male was observed to complete few matings and sired few or no chicks.

In a later study, White Leghorn flocks were kept and observed under crowded conditions more comparable to those seen in present-day hatchery flocks (Craig et al., 1977). In those flocks, the number of completed matings was not significantly associated with a male's social status, although there was a tendency for higher-ranking males to mate more. Special attention was directed toward recording male-to-male interferences with matings, but those were infrequent, a different finding from that reported by Guhl and his associates. It was suggested that the more crowded flocks and the presence of tube-type feeders might have reduced visibility of matings and prevented ready access to the mating pair by other males, as contrasted with the unobstructed view and ease of movement in the earlier studies. It also appeared likely that competition for access to a small number of hens may have been an important factor in increasing agonistic activity and interferences in the study of Guhl et al. (1945), where there were less than two per male as compared to the later experiment where each male was exposed to 8, 12, or 24 hens.

A further study was carried out by Kratzer and Craig (1980). Because group size and density differed so much in the previous studies, this experiment compared effects of social status of cocks on mating behavior in smaller and larger flocks and in high- and low-density pens. The effect of male social status was essentially the same whether the chickens were crowded or not and whether the number of hens was 30 to 60. Figure 17-3 shows results for the first 5 weeks after males were added to the flocks. Clearly, social rank did influence frequency of mounting and completed matings, and dominant males interfered more frequently with other males' mating attempts than did subordinate males. Nevertheless, subordinate and intermediate-status cocks, comprising 67% of the males present, accounted for 58% of total completed matings. It was also observed that the frequency of interferences with mating dropped off rapidly over the 5 weeks of observation, and attempted interferences were largely ignored after the first 3 weeks.

Sheep

Effects of male social dominance on mating frequency of rams has been investigated in a number of studies. The results obtained indicate that subordinates are likely to be greatly inhibited in sexual activity in relatively small pens and particularly when yearling rams are subordinate to mature rams and the number of ewes in estrus is low (Hulet et al., 1962). Dominance relationships were determined quickly and little fighting occurred thereafter. Dominant rams in multiple-sire pens did not differ in frequency of sexual acts (teases, mounts, and completed matings) as compared to rams kept singly in comparable pens. Also, dominant rams controlled nearly all mating activity in multiple-sire pens. However, when more than one ewe was in heat, the dominant male was unable to completely inhibit mating of his subordinates (especially when three males were present). Mating attempts by subordinates were frequently interrupted by the dominant ram. Under such conditions a dominant ram, if partially or completely sterile, could markedly reduce the lamb crop.

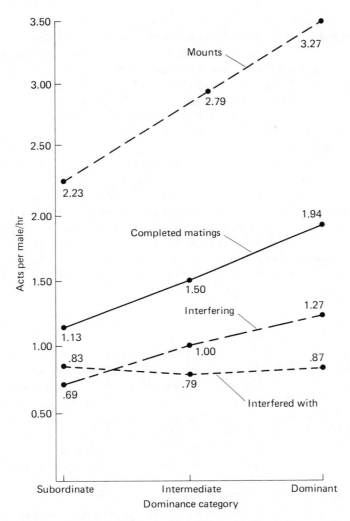

Figure 17-3 Effects of male dominance category on mating behavior and interferences with mating of other males in White Leghorn flocks. (From Kratzer and Craig, 1980.)

Lindsay et al. (1976) found that submissive rams mounted and mated estrous ewes fewer than one-third as many times when close to dominant rams, even though physical contact between rams was prevented, as compared to their performance in the absence of dominant rams. Dominant rams, under similar circumstances, were not influenced in their sexual behavior by the presence of subordinate rams.

In contrast to the results reported above, dominance status of rams has little, if any, effect on the frequency of mating when sheep are kept under extensive management conditions and when larger numbers of ewes are in estrus (Mattner et al., 1967, 1973; Fowler and Jenkins, 1976). An interesting and relevant observation is that "harems" of estrous ewes form around the dominant ram, compete for his attention, and probably reduce his tendency to interfere with subordinate rams, which are then free to search the remainder of the flock for receptive females (Mattner et al., 1967). Because it is known that there is an optimal time during estrus when matings are mostly likely to result in conception (Dzuik, 1970), it has been suggested (Fowler and Jenkins, 1976) that dominant rams may have more ready access to ewes at that time. Perhaps ewes join the harem at the time when conception is most likely. If that happens, subordinate rams would be mating with ewes during suboptimal parts of their estrous period. If this scenario is correct, dominant rams would sire more progeny in large flocks even though subordinate males mated as frequently.

Cattle

Multiple-sire breeding groups in cattle made up of bulls of widely differing ages are likely to exhibit social dominance effects on mating frequency. Osterhoff (cited by Blockey, 1978) used blood-typing information in determining paternity in a South African herd over a 5-year period. Three or four bulls were present each year in a herd. Older bulls were probably socially dominant; they clearly inhibited the sexual effectiveness of younger ones, as shown by 63 to 76% of all calves being sired by the oldest or second-oldest bull (Table 17-3).

Blockey (1978) compared breeding activity in four herds of heifers; two herds included one 5-year-old and two 2-year-olds, the other two had three 2-year-old bulls. Heifers in estrus spent 90% of their time in sexually active groups within each herd. The dominant male also spent over 90% of his time with this sexually receptive group. Subordinate males were in or close to the group just over half of the time. Subordinates were usually able to engage in matings when more than three heifers were in heat simultaneously. However, with three or fewer heifers in the receptive group, the older and dominant bull interrupted nearly nine-tenths of sexual "episodes" of his subordinates as compared to only one-fifth interruptions by dominant 2-year-old bulls in the young-bull mating groups. Apparently because of the unfavorable social-dominance effects in the mixed-age

Table 17-3 *Association of age of bull and proportion of calf crop sired by each bull in a group mating situation*

Bull's name	Age of bull and percentage of calves sired, by years									
	1964		1965		1966		1967		1968	
	Age	Calves sired	Age	Calves sired	Age	Calves sired	Age	Calves sired	Age	Calves sired
Oubass	10	70	11	76	12	12	Absent[a]		Absent[a]	
Matie	4	17	5	18	6	63	7	73	8	25
Morena	3	7	4	6	5	12	6	13	7	63
Slinger	2	6	Absent[a]		4	12	5	15	6	12

[a]Bull was absent from the herd.
SOURCE: Based on blood-typing information of Osterhoff, cited by Blockey (1978).

groups, conception rate during the second estrus period was about 64%, as compared to 81% in the young-male groups. Low fertility, if present in a dominant bull, could seriously delay the calf crop even if subordinate bulls were fully fertile.

Unlike subordinate rams, which in large pastures are able to mate at as high frequencies as the dominant one, subordinate bulls contribute much less even under extensive conditions. Thus, Rupp et al. (1977), with herds of about 100 heifers and four bulls per herd in large pastures (450 ha or 1120 acres) found that more heifers were marked by the dominant bull than by subordinate ones and that the number of marks generally varied according to dominance ranking. (Bulls were fitted with a chin-ball marking harness containing different colors to monitor mounting activity.) All bulls in a pasture associated with the single group of sexually active heifers.

Other Species

There is a dearth of information on social-dominance effects on sexual behavior of stallions kept together in relative small areas with receptive females. Under natural conditions, feral horses usually form into small bands, including an adult dominant male, a few mares, and their sexually immature progeny. Breeding activity of males is therefore controlled by the herd stallion and excess males are found in bachelor herds.

Adult boars, when first penned together as strangers, fight vigorously to establish dominance, which is usually determined within the first hour. As indicated by Signoret et al. (1975), "the dominant boar is seldom satisfied until he has pursued the loser and will continue to bite and slash the retreating animal. The loser backs away . . . and, if pursued, will turn and run. On subsequent meetings, the victor need only emit a warning glance, or a short grunt in order to induce submission." It seems likely that dominance so vigorously established would carry with it breeding rights if sexually receptive females were present.

The profound effect of large genetic differences in social-dominance ability on breeding activity of male dogs is apparent from a small-scale comparison of Terriers and Beagles reared in the same litters (James, 1951). Although raised as littermates, all Terriers were dominant to all Beagles, although adult Beagles are more than a third again heavier than Terriers. The dominant Terrier male of each group sired all puppies.

Summary

Low social status of males in multiple-male mating situations is likely to be associated with inhibition of mating behavior or interference by more dominant males when (1) males differ considerably in size, strength, or aggressive tendencies (as when Terrier and Beagle males compete for mating privileges); (2) there are relatively few sexually receptive females present at any one time; (3) the group is in a restricted area, so that dominant males are in the close presence and can easily interfere with subordinates; and (4) estrous females are in a sexually active group in close proximity to the dominant male(s). On the other hand, low-status males may make a significant contribution to flock or herd fertility in the absence of one or more of the conditions listed above.

THE BULLER-STEER SYNDROME

Cattle producers have known for many years that some steers, known as *bullers*, will stand for mounting by others. When kept on pasture or in small lots, the buller-steer syndrome does not usually pose a serious problem. However, under commercial feedlot conditions the incidence of bulling has increased and serious economic losses occur. This abnormal mating behavior differs from that seen in all-female groups, as bullers are unlikely to mount others that also stand (Brower and Kiracofe, 1978). It also differs from the situation seen when bulls and cows are kept together, as a dominant bull ordinarily inhibits or drives away other bulls interested in a sexually receptive cow, but "rider" steers apparently take turns mounting bullers.

Because of persistent mounting, buller steers may be injured or become exhausted and may collapse and die unless removed to a separate lot.

Why the incidence of the buller syndrome has risen so much in recent years is not entirely clear, but it is apparent that many stressors are acting, especially when cattle are first introduced into large feed-lots. They are likely to be tired and frightened, they are crowded, many strange animals are present, and they are handled several times (vaccinated, dipped, implanted with hormone pellets, etc.). Among several measures taken, Brower and Kiracofe (1978) detected ele-vated levels of adrenal corticoids in the urine and higher total estro-gens in the serum of bullers as compared to normal steers. However, Irwin et al. (1979) reported that both serum estradiol and testo-sterone were lower while steers were being mounted than later, after recovery. Thus, the physiological status of bullers relative to circulat-ing hormones remains obscure. Interestingly, buller steers are be-lieved to be fast and efficient gainers when protected from mounting activity.

Because the buller syndrome occurs at relatively low frequencies, is erratic in occurrence, involves valuable animals, and may be brought on by a variety of causes, it does not lend itself to small-scale, con-trolled experiments. Nevertheless, carefully managed commercial cattle feedlots have supplied some valuable information. All recent studies conducted on feedlot steers involved animals receiving estrogenic hormones in one form or another, and in some cases different hormone treatments were applied to different groups within years or over several years.

A very large-scale study by Pierson et al. (1976) led them to conclude that there was a definite relationship between hormone implantation and the frequency of bulling and that the incidence had been increasing. Thus, the average for the period 1968–1970 was 1.5% and for 1971–1974 it was nearly 3%; during the first 3 years a synthetic estrogen was fed, but for the later years the level of hor-mone in the feed was increased and three different anabolic hormones were implanted as well. In the final year, Pierson et al. found that not only was the estrogenic compound fed at a high level, but some cattle fed for more than 60 days were reimplanted with an additional hormone pellet; the incidence of bulling was nearly 4% in that year. Reimplantation of hormone pellets for steers kept on feed over long periods is becoming more common and Schake et al. (1979) found a second wave of bullers following reimplantation 91 days after the initial treatment; the percentage of bullers from the initial treatment was about 1% and an additional 2½% occurred after reimplantation. Hormone pellets including both estrogen and progesterone increased the incidence of the buller syndrome over that occurring when the pellets included a synthetic estrogen only (Pierson et al., 1976; Irwin et al., 1979; Schake et al., 1979).

Although the evidence from feedlot studies implicates hormone growth stimulants as one cause of the buller syndrome, it is also clear that other factors influence its incidence as well. Thus, Schake et al. (1979) found that cattle from two different sources, although in the same feedlot and treated alike, differed significantly in the incidence of bullers; steers from one source had over three times as many bullers as those from the other source. The factor or factors causing this difference could not be identified as animals from the two sources had been previously exposed to many different variables and may have represented quite different genetic stocks (they were all crossbreds). Among variables (other than hormone treatment) identified with some certainty as predisposing steers to the buller syndrome are the keeping of larger numbers per group and the repeated introduction of strangers into lots (Irwin et al., 1979).

In most large feedlots steers are observed closely at least once a day so that steers in the process of becoming bullers may be separated into a separate lot before serious injuries occur. Although this method is relatively effective, it increases the complexity of management required and involves additional labor and other costs.

REFERENCES

ALEXANDER, G., J. P. SIGNORET, and E. S. E. HAFEZ, 1974. Sexual and maternal behavior. *In* E. S. E. Hafez (Ed.), Reproduction in Farm Animals, 3rd ed. Lea & Febiger, Philadelphia.

ALMQUIST, J. O., and E. B. HALE, 1956. An approach to the measurement of sexual behaviour and semen production of dairy bulls. Proc. 3rd. Int. Cong. Anim. Reprod., Camb., Plenary Pap., pp. 50–59.

BEACH, F. A., 1968. Coital behavior in dogs. III. Effects of early isolation on mating in dogs. Behaviour 30: 217–237.

BEILHARZ, R. G., 1978. Effect of sexual behaviour on reproductive success. Proc., 1st World Congr. Ethol. Appl. Zootech., Madrid, pp. 503–506.

BLOCKEY, M. A. de B., 1976. Sexual behaviour of bulls at pasture: a review. Theriogenology 6: 387–392.

——, 1978. Serving capacity and social dominance of bulls in relation to fertility. Proc. 1st World Congr. Ethol. Appl. Zootech., Madrid, pp. 523–530.

BROWER, G. R., and G. H. KIRACOFE, 1978. Factors associated with the buller-steer syndrome. J. Anim. Sci. 46: 26–31.

CRAIG, J. V., and A. L. BHAGWAT, 1974. Agonistic and mating behavior of adult chickens modified by social and physical environments. Appl. Anim. Ethol. 1:57-65.

CRAIG, J. V., B. AL-RAWI, and D. D. KRATZER, 1977. Social status and sex ratio effects on mating frequency of cockerels. Poult. Sci. 56:767-772.

DZUIK, P., 1970. Estimation of optimum time for insemination of gilts and ewes by double-mating at certain times relative to ovulation. J. Reprod. Fertil. 22:277-282.

FOOTE, R. H., 1974. Artificial insemination. *In* E. S. E. Hafez (Ed.), Reproduction in Farm Animals, 3rd ed. Lea & Febiger, Philadelphia.

FOWLER, D. G., and L. D. JENKINS, 1976. The effects of dominance and infertility of rams on reproductive performance. Appl. Anim. Ethol. 2:327-337.

FOX, M. W., 1975. The behaviour of cats. *In* E. S. E. Hafez (Ed.), The Behaviour of Domestic Animals, 3rd ed. The Williams & Wilkins Company, Baltimore.

——, and M. BEKOFF, 1975. The behaviour of dogs. *In* E. S. E. Hafez (Ed.), The Behaviour of Domestic Animals, 3rd ed. The Williams & Wilkins Company, Baltimore.

FRASER, A. F., 1968. Reproductive Behaviour in Ungulates. Academic Press, Inc., New York.

GROSSE, A. E., and J. V. CRAIG, 1960. Sexual maturity of males representing twelve strains of six breeds of chickens. Poult. Sci. 39:164-172.

GUHL, A. M., and D. C. WARREN, 1946. Number of offspring sired by cockerels related to social dominance in chickens. Poult. Sci. 25:460-472.

GUHL, A. M., N. E. COLLIAS, and W. C. ALLEE, 1945. Mating behavior and the social hierarchy in small flocks of White Leghorns. Physiol. Zool. 18:365-390.

HALE, E. B., 1955. Defects in sexual behavior as factors affecting fertility in turkeys. Poult. Sci. 34:1059-1067.

HARLOW, H. F., 1965. Sexual behavior in the Rhesus monkey. *In* F. A. Beach (Ed.), Sex and Behavior. John Wiley & Sons, Inc., New York.

HEMSWORTH, P. H., R. G. BEILHARZ, and D. B. GALLOWAY, 1977. Influence of social conditions during rearing on the sexual behaviour of the domestic boar. Anim. Prod. 24:245-251.

HULET, C. V., S. K. ERCANBRACK, R. L. BLACKWELL, D. A. PRICE, and L. O. WILSON, 1962. Mating behavior of the ram in the multi-sire pen. J. Anim. Sci. 21:865-869.

IRWIN, M. R., D. R. MELENDY, M. S. AMOSS, and D. P. HUTCHESON, 1979. J. Am. Vet. Med. Assoc. 174:367-370.

JAMES, W. T., 1951. Social organization among dogs of different temperaments, Terriers and Beagles, reared together. J. Comp. Physiol. Psychol. 44:71-77.

JUSTICE, W. P., G. R. McDANIEL, and J. V. CRAIG, 1962. Techniques for measuring sexual effectiveness in male chickens. Poult. Sci. 41:732-739.

KRATZER, D. D., and J. V. CRAIG, 1980. Mating behavior of cockerels: Effects of social status, group size, and group density. Appl. Anim. Ethol. 6:49-62.

LIGHTFOOT, R. J., and J. A. C. SMITH, 1968. Studies on the number of ewes joined per ram for flock matings under paddock conditions. 1. Mating behaviour and fertility. Aust. J. Agric. Res. 19:1029-1042.

LINDSAY, D. R., D. G. DUNSMORE, J. D. WILLIAMS, and G. J. SYME, 1976. Audience effects on the mating behaviour of rams. Anim. Behav. 24:818-821.

MATTNER, P. E., A. W. H. BRADEN, and K. E. TURNBULL, 1967. Studies in flock mating of sheep. 1. Mating behaviour. Aust. J. Exp. Agric. Anim. Husb. 7:103-109.

MATTNER, P. E., A. W. H. BRADEN, and J. M. GEORGE, 1973. Studies in flock mating in sheep. 5. Incidence, duration and effect on flock fertility of initial sexual inactivity in young rams. Aust. J. Exp. Agric. Anim. Husb. 13:35-41.

McDANIEL, G. R., and J. V. CRAIG, 1959. Behavior traits, semen measurements and fertility of White Leghorn males. Poult. Sci. 38:1005-1014.

OSBORNE, H. G., L. G. WILLIAMS, and D. B. GALLOWAY, 1971. A test for libido and serving ability in beef bulls. Aust. Vet. J. 47:465-467.

PARKER, J. E., F. F. MCKENZIE, and H. L. KEMPSTER, 1940. Observations on the sexual behavior of New Hampshire males. Poult. Sci. 19:191-197.

PEPELKO, W. E., and M. T. CLEGG, 1964. Factors affecting recovery of sex drive in the sexually exhausted male sheep (*Ovis aries*). Fed. Proc. Fed. Am. Soc. Exp. Biol. 23:362.

PEXTON, J. E., and P. J. CHENOWETH, 1977. Using bulls to breed beef heifers at a synchronized estrus. Proc. 69th An. Meet. Am. Soc. Anim. Sci., Madison, Wis., p. 195 (Abstract).

PIERSON, R. E., R. JENSEN, P. M. BRADDY, D. P. HORTON, and R. M. CHRISTIE, 1976. Bulling among yearling feedlot steers. J. Am. Vet. Med. Assoc. 169:521-523.

RUPP, G. P., L. BALL, M. C. SHOOP, and P. J. CHENOWETH, 1977. Reproductive efficiency of bulls in natural service: effects of male to female ratio and single- vs. multiple-sire breeding groups. J. Am. Vet. Med. Assoc. 171:639-642.

SCHAKE, L. M., R. A. DIETRICH, M. L. THOMAS, L. D. VERMEDAHL, and R. L. BLISS, 1979. Performance of feedlot steers reimplanted with DES or Synovex-S. J. Anim. Sci. 49:324-329.

SCHEIN, M. W., and E. B. HALE, 1965. Stimuli eliciting sexual behavior. *In* F. A. Beach (Ed.), Sex and Behavior. John Wiley & Sons, Inc., New York.

SIEGEL, P. B., 1965. Genetics of behavior; selection for mating ability in chickens. Genetics 52:1269-1272.

——, and H. S. SIEGEL, 1964. Rearing methods and subsequent sexual behaviour of male chickens. Anim. Behav. 12:270-271.

SIGNORET, J. P., B. A. BALDWIN, D. FRASER, and E. S. E. HAFEZ, 1975. The behaviour of swine. *In* E. S. E. Hafez (Ed.), The Behaviour of Domestic Animals, 3rd ed. The Williams & Wilkins Company, Baltimore.

SMYTH, J. R., and A. T. LEIGHTON, 1953. A study of certain factors affecting fertility in the turkey. Poult. Sci. 32:1004-1013.

WILSON, E. R., R. K. JOHNSON, and R. P. WETTEMANN, 1977. Reproductive and testicular characteristics of purebred and crossbred boars. J. Anim. Sci. 44:939-947.

WOOD-GUSH, D. G. M., 1957. Aggression and sexual activity in the Brown Leghorn cock. Br. J. Anim. Behav. 5:1-6.

——, 1960. A study of sex drive of two strains of cockerels through three generations. Anim. Behav. 8:43-53.

——, 1963. The relationship between hormonally-induced sexual behaviour in male chicks and their adult sexual behaviour. Anim. Behav. 11:400-402.

18

Sexual Behavior of Females

MALE PRESENCE
AND SEXUAL MATURITY

The rate of female sexual development depends not only on seasonal and other environmental effects (both physical and social) as seen in earlier chapters, but the presence of adult males can also significantly accelerate sexual maturity in some species.

Birds

Among pair-bonding birds such as canaries and ring doves, the presence of an adult male stimulates estrogen production by the female and the onset of behaviors essential to reproduction (Hinde, 1965; Lehrman, 1965). Visual and auditory stimuli alone are enough to induce female domestic pigeons to lay eggs (Lehrman, 1965). The gallinaceous order, which includes most species of domesticated birds, appears to be exceptional in that ovarian development and egg laying appear to be independent of male stimulation. Certainly, domesticated turkeys and chickens lay large numbers of eggs even if kept away from males throughout their lifetimes. Hale et al. (1969:562) failed to find earlier maturity of turkey females that were courted and mated by males as compared with those that were isolated; however, it is not clear whether their isolation involved complete absence of male stimuli, beyond physical absence from pens. Similarly, there was no apparent effect of males being kept with young female chickens on age at maturity or rate of egg production (Bhagwat and Craig,

1979). However, all-female flocks were exposed to the presence of males in neighboring flocks; therefore, the effects of complete male absence from the hen house was not tested.

Mammals

In his book on reproductive behavior in ungulates, Fraser (1968:58–60) points out that Australian researchers working with sheep in the early 1950s found evidence that the addition of a ram to a flock of ewes stimulated their reproductive behavior. For example, Schinckel (1954) found that when a ram was added to a flock of ewes shortly before the regular breeding season, the average onset of estrus was about 2 weeks earlier than in the control flock and there was a high incidence of estrus synchrony, with most ewes coming into heat 15 to 17 days later. It soon became apparent that at least some of the effects of the ram could be obtained even if he was not in immediate physical contact with the ewes. Of sight, sound, and odor, it now seems that odor must be most important, so that pheromones are implicated.

Pheromone Effects. Pig producers are interested in the early onset of puberty because less time, labor, and feed are required before first litters are produced, or if gilts are allowed extra cycles before breeding, larger litters may be expected. Several studies indicate that placing mature boars in the presence of young gilts causes them to reach puberty up to 30 to 40 days earlier than would occur otherwise.

That the stimulating effect of a boar may be seasonally dependent is suggested by the results of Mavrogenis and Robison (1976), as shown in Table 18-1. Their spring-farrowed gilts, placed in a pen

Table 18-1   *Effects of male presence and season of birth on age (in days) at first detectable estrus in gilts*

Season when treated gilts were farrowed	Treatment		Difference
	Isolated from boar	Boar in next pen (from 140 days of age)	Isolated minus boar present
Spring	225	199	26[a]
Fall	203	199	4[b]

[a]Probability of difference due to chance is less than 1%.
[b]Not significant.
SOURCE: From results presented by Mavrogenis and Robison (1976).

next to a boar, showed signs of estrus (i.e., red and swollen vulvas and standing response to pressure on the hips) 26 days earlier than those isolated from boars. On the other hand, fall-farrowed gilts did not mature earlier when a boar was in the next pen. All gilts were slaughtered within 10 days of showing signs of puberty and their ovaries were checked to determine whether eggs had been shed in an earlier cycle without overt signs of estrus. Gilts exposed to boars had more first-cycle ovulations without showing estrus. Thus, first ovulations actually occurred earlier, on the average, when a boar was present in an adjacent pen than the 199 days indicated in Table 18-1. In a second experiment involving spring-farrowed gilts only, the presence of a boar in an adjacent pen hastened age at first estrus by 40 days.

Studies with mice in laboratory situations have yielded results that are roughly parallel to those obtained with swine. Because of the precise control of environmental variables possible in the laboratory, the importance of adult male pheromones in accelerating the onset of first estrus in mice has been established beyond reasonable doubt. The juvenile female mouse is especially sensitive to male pheromones at certain ages, and stimulation during a brief sensitive period may be as effective as prolonged stimulation (Colby and Vandenbergh, 1974).

Additional findings with laboratory mice are suggestive of new avenues of research that might be of considerable importance to livestock producers. Thus, Colby and Vandenbergh, although able to advance first estrus by 4 to 6 days using adult male pheromone treatment only, did not obtain as great an advance in maturity as in an earlier study (Vandenbergh, 1967) in which adult males were caged with groups of weanling and older female mice; direct exposure from weaning resulted in a 20-day advance in maturity. Exposure to pheromones produced by adult virgin females had an opposite effect: it retarded the rate of sexual development of immature females (Colby and Vandenbergh, 1974).

PROCEPTIVITY AND RECEPTIVITY

Proceptivity refers to sexual appetitive behavior of the female, and *receptivity* is the consummatory phase, in which the female assumes a mating stance allowing the male's mounting, intromission, and intravaginal ejaculation (Beach, 1976). A female's appetitive or proceptive behavior may be taken as a measure of how attractive a male is to her. Because of the mutual stimulatory effect, a male may behave in ways that make him more attractive when an estrous female is present. For example, boars have a mating chant that is heard only when they are sexually excited. On the other hand, they secrete phero-mones constantly.

In considering female proceptive behavior, it is of interest to know about the components of male attractiveness. Are there simple releasers? If so, they might be used to determine the phase of the female's reproductive cycle in the absence of a male. Such informa-tion would be valuable for those using artificial insemination in mam-mals. Some cows, sows, and mares are difficult to detect in heat, and ewes present a difficult problem because they are only rarely mounted by other females when in heat and show no behavioral signs that are obvious to man. Nevertheless, as Lindsay and Fletcher (1972) have shown, ewes show considerable proceptive behavior; 75% of those in heat will come to a tethered ram and 100% will come if they are given high dosages of estrogen.

Signoret's Studies with Swine

The French investigator Signoret has carried out a series of studies with swine that have increased understanding of attractivity of the boar to the sow, and vice versa (Signoret, 1970a, 1971; Signoret et al., 1975). Although Signoret's studies give considerable insight into the behavior of swine, they do not necessarily indicate what the situ-ation will be in other species. As an example, the sow kept in an all-female group becomes nervous and moves about much more for several days before and during estrus, as if searching for a male (Sig-noret et al., 1975), but the ewe shows no increase in activity in the absence of a ram (Lindsay and Fletcher, 1972). In the same way, diestrous sows nose the flanks of estrous females, attempt to mount, and may do so; among ewes there is little interest shown in the estrous female by other ewes.

Proceptive Behavior. Sows, unlike boars, readily discriminate between appropriate and inappropriate sexual partners. This discrim-inative ability and associated proceptive behavior are evident for about a day before, during, and 2 days after standing heat. Signoret used a T-maze apparatus in which contrasting animals (male vs. fe-male, intact male vs. castrated male, etc.) were confined in the two arms and then measured the appetitive behavior of sows in terms of the amount of time spent near each of the two. While in estrus, sows spent over four times as much time near intact males as they did near castrated ones; during diestrus they divided their time equally. Even when anesthetized and not visible, intact boars were more attractive than castrated ones.

On the basis of an isolation study, Signoret (1970b) suggested that appetitive and receptive behaviors of the estrous gilt toward boars were instinctive. Gilts were placed in individual isolation shortly after birth and kept physically and visually separated (gilts did see each other occasionally) until about a year of age. When tested in a T-maze, anestrous gilts were equally attracted by a boar or another gilt, but gilts in heat spent more time near the boar. Essentially the same results were obtained as with normally reared gilts.

Receptive Behavior. Behavioral estrus begins (by definition) when the female assumes the mating stance so that the male may mount and copulate. When estrous sows receive pressure on their backs from a human being, a certain proportion will show the typical

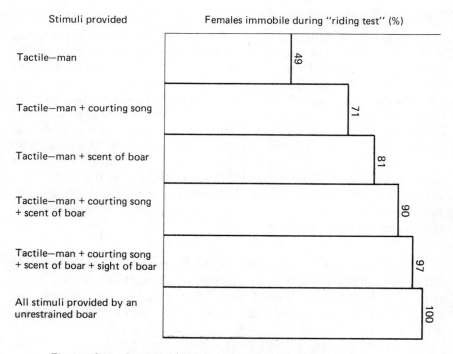

Figure 18-1 Stimuli effective in causing the immobility or standing reaction of sows to the "riding test" for females that would stand to be mounted by a boar. (Drafted from results presented by Signoret, 1971.)

immobility or standing response so that a person can sit astride. The "riding test" (by man) causes, on the average, only about half of females to stand that would assume the mating stance in response to the presence of a boar; the incidence of standing for the riding test varies over time during the estrous period, but even at the optimal time is no higher than 60% (Signoret, 1971). Stimuli provided by the boar increase the sow's willingness to become immobile during the riding test, as shown in Figure 18-1.

These results provide for some fascinating practical applications, particularly because the courting song of the boar is effective when recorded and played back and because synthetically produced boar pheromones are now available in aerosol-spray form. A swine producer wishing to use artificial insemination could theoretically dispense with the use of boars for detecting estrus and expect to identify 90% of sows ready for insemination by using the riding test along with a recorded courting song and an aerosol spray of boar odor (Figure 18-1). Hillyer (1976) used synthetically produced boar odor in accelerating the return of sows to estrus following weaning of pigs at 5 weeks of age. His sows which were not sprayed on the nose with boar odor required nearly 4 weeks before returning to heat and allowing natural mating, but those sprayed either once or twice, shortly after weaning, returned more rapidly. Because of the pheromone-spray treatment, the interval between weaning and conception was reduced by more than 2 weeks. There was no reduction in litter size resulting from the accelerated return to heat in the treated sows.

Male Presence and Duration of Estrus

Two carefully conducted studies with sheep have established that the continuous physical presence of rams dramatically reduces the length of estrus as compared to its duration when ewes are "teased" (i.e., allowed a brief contact with rams at 4-hour intervals; Table 18-2). Because ewes cannot be detected as being sexually receptive without being checked for the immobility response by a ram, it is not known whether they remain potentially receptive over a longer period if no male is present. Physical contact with a ram, even without mounting activity, was found to reduce the estrous period significantly (Table 18-2). However, the combined effects of pheromones, auditory, and visual stimuli associated with a ram do *not* reduce the length of estrus; Parson and Hunter (1967) kept rams in adjacent pens to estrous ewes without reducing the duration of the heat period.

It is of interest to note from the table that the percentage reduc-

Table 18-2 *Duration of estrus of ewes as affected by continuous presence of rams vs. teasing at 4-hour intervals*

Item	Three-month averages[a]	Experiment 1[b]
Status of ewes	Intact	Ovaries removed
Cause of estrus	Natural	Hormonal
Duration, hours		
When teased at 4-hr intervals	24.3	40.1
Continuous presence of unrestricted ram(s)	13.5 (56%)[c]	21.2 (53%)[c]
Continuous presence, ram unable to mount	15.3 (63%)[c]	—

[a]Data from Parson and Hunter (1967).
[b]Data from Fletcher and Lindsay (1971).
[c]Differs significantly from 4-hr teasing.

tion in duration of the heat period is essentially the same when rams are kept with either intact ewes undergoing natural estrus or with ovariectomized females in which heat has been induced by hormone treatment. From this evidence, Fletcher and Lindsay (1971) argued that there is inhibition of the neural mechanism controlling the expression of estrus when male and female sheep are in continuous association. A possibly related finding is that mating hastens the onset of ovulation within the estrous period and reduces the time required for ovulation to occur in the sow (Signoret et al., 1972).

Preferential Mating

It seems likely that preferential mating exists in all domesticated species of animals, but its relative importance varies considerably. Thus, female dogs appear to be particularly choosy about mates, but sows in heat usually accept any boar readily. Both male and female are obviously involved and, when competition for mates exists, socially dominant individuals tend to succeed in copulation while others are either driven away, inhibited from attempting copulation, or interfered with during mating. But individual preferences will now be considered when competition based on social rank is not involved.

Chickens. When males of the same breed are placed singly with flocks of hens, sexual crouching by the hens and number of completed matings may differ greatly from male to male. Preferences of individual hens for particular males tend to be like those of the flock as a whole (Lill and Wood-Gush, 1965). In attempting to determine whether certain elements of males' courtship behavior were more effective in eliciting female receptivity, Lill and Wood-Gush identified the waltz and rear approach as the only two (of 13 examined) that consistently preceded crouching to a significant degree. Two of six males tested, eliciting the least female crouching, did not approach from the rear and differed from all other males in that respect. However, it was also apparent that more than courting behavior of a particular male was involved in his popularity; some males received crouches on entering a pen, even before exhibiting courtship displays.

Brown Leghorn hens of a particular line were tested by Lill and Wood-Gush (1965) for sexual crouching and copulation by introducing cockerels, one at a time, from their own line and from two other Brown Leghorn lines that differed in plumage color, body size, conformation, and other characteristics. Hens crouched much more frequently for males of their own line and a large number of matings were completed with them while not a single copulation occurred with males of the other lines. The tests were brief (15 minutes) and it may be asked whether "anxiety" generated by the different appearance and strange characteristics of other-line males may not have contributed to their lack of success. In a later study, Lill (1968a) found that temporary introduction of White Leghorn hens into Brown Leghorn female flocks resulted in mutual avoidance lasting over 15 minutes, on the average.

That homogamy, or mating of like with like, is related to experience was established by Lill (1968b). When Brown Leghorn hens were reared from hatching with cockerels of their own strain, they strongly preferred them in brief mating tests over White Leghorn males. However, when Brown Leghorn hens were reared from hatching in a flock with White Leghorn males, they showed no discrimination; crouching was elicited just as frequently by either Brown or White Leghorn cockerels.

Commercial hatcheries produce crossbred and straincross chicks in very large numbers, and in some of those crosses the hens and cocks differ greatly in plumage color, size, conformation, and other characteristics (as when White Leghorn and Rhode Island Reds are crossed). Those matings obviously succeed, although mating behavior may be inhibited initially, as in the studies of Lill and Wood-Gush. Nevertheless, these and other experimental results suggest that when breeds or strains are to be crossed, better fertility may result if pullet and cockerel chicks of the stocks to be crossed are reared together.

Wild and Game-Farm Ducks. When game-farm birds of any species are released in large numbers, there may be some risk that "unfit" genes from the semidomesticated stock may enter into the wild population gene pool at a significant level. This consideration led Cheng et al. (1979) to look into mate preferences of wild and game-farm Mallard ducks when reared with their own kind or with the other stock. Equal numbers of males and females of the different stock and rearing combinations were placed in seminatural outdoor pens. Wild drakes were somewhat suppressed in their sexual behavior because the game-farm males were aggressive and dominant to them. However, unpaired wild females were also likely to seek cover or stay along fence edges or in corners along with the wild males. In this situation game-farm males reared with their own strain of females paired with them only. In the same way, wild drakes reared with wild-stock females paired only with wild females. However, game-farm males reared with wild females, and wild males reared with game-farm females, were successful in pairing with either kind of female.

Cats and Dogs. After examining the results of many mating tests among all combinations of five male and three female cats, Whalen (1963) suggested that there were male–female interactions. Although intromissions were achieved in all combinations, some pairings gave fewer and others more matings than expected, based on the average performance of the same individuals with other animals.

When Beagle males are tethered and females are allowed to rove freely, it becomes clear that female dogs enjoy considerable freedom from the effects of their ovarian hormones (LeBoeuf, 1967). Although bitches only copulate when estrogen levels are high, they show marked preferences and aversions in their choice of males. Thus, a female dog in the reproductive phase (about to ovulate) may be receptive to one male but may aggressively rebuff another. Some male differences were also exhibited when they were allowed freedom of movement and females were tethered, but those differences were not so large. Female preferences did not seem to be related to the dominance relationships among the males, as one of the two most aggressive males was readily accepted by all females, whereas the other was not.

Sheep. When the same rams were allowed freedom of movement or were tethered in lots containing estrous ewes it was learned that those rams mating more frequently, when free to move about, also mated more frequently when tethered. From this observation, Lindsay and Robinson (1961) reasoned that ewes were attracted to the more vigorous ram. Rams were of three breeds, but the crossbred ewes did not appear to show any decided preference for one breed over the other when given a free choice of tethered rams. Details of their rearing environments are not given, so that the significance of that observation cannot be judged.

Some Generalities

Appetitive sexual behavior or proceptivity by the female, although not so obvious as the sex drive of the male, is evident across a wide variety of species. Thus, if not sought out by the male at the appropriate time in her reproductive cycle, the female will approach the male so that mating can occur. Elements of male attractivity have been investigated especially in swine, and it is apparent that the courting chant and boar pheromone contribute significantly to male attractiveness. Boar pheromone is also associated with an earlier return to sexual receptivity of the sow after weaning a litter.

Physical contact with a male reduces the duration of estrus, at least in some mammalian species. Receptivity toward particular individuals varies, resulting in preferential mating, which is more important in some species (e.g., cats and dogs) than in others. In chickens and ducks, being reared with males of particular genetic stocks increases the likelihood of preferential mating with other males of the same stock.

FEMALE ATTRACTIVITY

Sexual attractivity of the female is measured in terms of male response. When a sexually active male and a receptive female are placed together under favorable circumstances (the environment is familiar, etc.), a mating will usually occur promptly. A refractory interval then intervenes, the length of which is fairly typical, depending on the species and the characteristics of the individuals involved. With apparent continuing receptivity on the part of the female, repeated matings are likely to be separated by longer and longer refractory periods until copulation stops. If at that time a different female is placed with the male, copulation may be resumed promptly and refractory periods will again be shortened. It is evident, then, that the increasing length of intervals between matings is not caused entirely by fatigue of the male, but that the original female, for some reason, becomes less attractive.

Almquist and Hale, working at Pennsylvania State University, were particularly effective in exploring factors that stimulate bulls to repeated ejaculations and short time intervals between ejaculations (see Hafez et al., 1969). Among important variables in maintaining high stimulus pressure is the animal being mounted. Surprisingly, after a bull or a ram has become satiated, he can be aroused by removing and returning the *same* female, or if the female is restrained within a pen, by moving it to a different location only a few feet away. Nevertheless, a different female, especially a different female that has not recently been mated by another male, is a more stimulating sexual object for returning mating activity to its initial high level; Pepelko and Clegg (1965) demonstrated this convincingly with sheep. Although artificial insemination centers commonly collect semen from bulls only once or twice a week, Hale and Almquist have shown that daily collections are possible without ill effects and that under optimal conditions, including frequent changes of stimulus animals, as many as 77 ejaculations can be obtained in a 6-hour period. However, because sperm reserves are rapidly depleted under such conditions, practical collection schedules would not make use of such high levels of sex drive.

Evidence that the same situation prevails in swine was obtained by Jakway and Sumption (in Hafez et al., 1962). When rested boars were placed in a pen containing a number of receptive sows, they mated with average frequencies of eight ejaculations in 135 minutes. However, in pair matings with single receptive females, boars did not mate a second time during 40-minute observation periods. Further testing showed that presentation of a different estrous female after each ejaculation was effective in inducing further activity, although the time interval gradually increased until a boar would stop mating for prolonged periods.

ESTRUS DETECTION

By Males of the Species

When a cow comes into estrus and no bull is nearby, she becomes hyperactive in a predictable way. She responds to stimuli ordinarily ignored and walks about a great deal, temporarily disregarding the niceties of social status by indiscriminately approaching dominant and subordinate herdmates so that agonistic interactions increase in frequency (Schein and Fohrman, 1955), she bellows, she attempts to mount other cows, and stands when mounted by them. In addition, there are external physiological indicators of her state; her vulva becomes swollen and there may be a mucous discharge, and she may emit pheromones in her urine or by other glandular secretions. In sizable groups of nonpregnant cows, it is common to see a sexually active female group in which the cows move about together and the signals indicated are given off.

It appears that when a sexually active group of cows is present, a bull is drawn to it primarily by the cow-to-cow mounting activity rather than by olfactory cues (Mattner et al., 1974). At close range a bull will sniff and lick the vulva, rest his chin on the cow's rump, and mount if she stands (Figure 18-2). It is obvious that different signals may be used at different distances and that tactile and gustatory stimuli are possible only after physical contact. Which indicators are of primary importance for identification of sexual receptivity and for providing sufficient stimulus, so that a male will mount and complete a mating? This question is appropriate for all species.

When artificial insemination is used, adequate identification of the estrous state may not be achieved because the human observer cannot read the signs as well as a male of the same species. However, infertility may also be a serious problem in natural matings. A few studies will be examined dealing with the importance of pheromones in the detection of estrus by males.

Pheromones. Chemical attractants put forth by the female in

Figure 18-2 A bull tests a heifer for willingness to be mounted. She moves away, indicating that standing estrus is not yet present. (Courtesy of L. R. Corah, Kansas State University.)

estrus appear to be of widely differing importance among mammalian species. Although swine are commonly thought of as relying heavily on the sense of smell, boars placed in T-mazes having estrous and anestrous sows in the two arms are only slightly more attracted to the estrous one (Signoret, 1970a). A different approach was used by Lindsay (1965) in working with sheep; he removed the olfactory sense of some rams surgically while leaving others intact and then compared their efficiency in finding estrous ewes after the anosmic rams had recovered from surgery. Although rams deprived of their sense of smell remained active in seeking out ewes, their efficiency was impaired; they approached anestrous ewes just as readily as they did estrous ones. On the other hand, intact rams approached estrous ewes 2¾ times more frequently than anestrous ones when equal numbers of the two kinds were present.

The powerful sense of smell in dogs is believed to be responsible for the accumulation of loose males about the homes of those having bitches in heat, even though those females are kept out of sight. (Of course, auditory stimuli are not excluded in such a case.) Cotton balls impregnated with vaginal secretions from estrous females receive more licking, chewing, and sniffing by males than those from anestrous bitches (Beach, 1976). That the urine may contain pheromones associated with estrus was shown by Dunbar (cited by Beach, 1976), who found that it was attractive to male dogs even when withdrawn directly from the bladder so as to be free of vaginal secretions.

The importance of pheromones for attracting the bull to the estrous cow is somewhat controversial. The lip curl or flehmen response of bulls (see Figure 5-1) as they sniff in the presence of estrous cows and while examining urine has been taken by many as indicating that odor must be important. Hart et al. (1946) indicated that olfactory cues were all-important in the arousal of sexual behavior, but Hafez and Bouissou (1975) questioned this on the basis that the results could just as easily be explained on the basis of visual stimuli. Hafez and Bouissou concluded, after reviewing other studies, that visual cues provided by estrous cows attract bulls far more effectively than do their odors.

Whether or not pheromones are important in attracting the bull to the estrous cow, their presence has been clearly established. Kiddy et al. (1978), concerned about failures in detecting estrus as a major cause of reproductive inefficiency with artificial insemination, showed that dogs can be trained to discriminate between estrous and diestrous cows by the sense of smell. After being trained with vaginal swabs and urine samples, the animals performed just as well in a farmyard setting where distractions were expected to affect performance adversely. These investigators compared samples from cows in full estrus with those taken about midway between heat periods. In comparing four dogs by various testing procedures, consistent differences were found; the best dog averaged 88% correct detections and the poorest about 70%.

General. That males of each mammalian species effectively detect the presence of sexually receptive females of their own kind is apparent, but the relative importance of the various cues that are available remains somewhat obscure. It seems reasonable to assume that pheromones of estrous females serve to broadcast the fact that they are present, but visual, auditory, or standing behavior may be required for final identification. The relative importance of pheromones seems to vary greatly among species. Thus, they appear to be important in dogs, relatively significant in sheep, and of uncertain importance in swine and cattle.

For Artificial Insemination in Cattle

The extensive use of artificial insemination in cattle and its increasing popularity in other species makes the accurate identification of the estrous period by man important. This is not a difficult task if a male can be used to tease the females into exhibiting typical mating stances in his presence. However, the necessity of keeping one or more males for the sole purpose of identifying females in estrus has little appeal to most livestock producers and may involve considerable expense, inconvenience, and an element of danger (as with bulls). However, cattle producers commonly use cow-to-cow mounting as an indicator (see Figure 16-1).

In groups of cows synchronized for ovulation by hormone treatment, it may be possible to achieve satisfactory conception rates by "inseminating by appointment," regardless of whether the cows show overt symptoms of heat (Lauderdale et al., 1974). Nevertheless, it is likely that many cattle producers will not use such a method, but will continue to look for signs of heat and have females bred artificially, using estrus as the criterion of when ovulation will occur.

Because of economic considerations, dairy cows should calve at intervals not exceeding 12 months. In practice, such calving intervals for entire herds are difficult to achieve, particularly when a bull is not present to aid in heat detection. Many contend that estrus is more difficult to detect under recently adopted management systems, especially when cows are kept continuously in confinement. Part of the problem is likely to be associated with the increased use of mechanical equipment, so that caretakers spend less time around females when estrus could be observed in the absence of feeding and milking distractions.

Silent Heats or Quiet Ovulations. Some cattle producers speak of "silent heats" when describing ovulations not accompanied by clear signs of sexual receptivity. That usage confuses behavioral estrus and the physiological event of ovulation. The terminology of "silent ovulation" would be more accurate. Some heats are silent only in the sense that the caretaker is absent or relatively inattentive when they occur.

Numerous reports (Kiracofe, 1980) indicate that uterine involution and ovulation commonly precede the first estrus following parturition. Dairy cows are likely to ovulate within 14 to 22 days after calving without showing estrus. Kiracofe's review indicates that uterine involution should permit conception by 6 weeks after giving birth in the cow and ewe or after 3 weeks in the sow, and that ova from those three species are seldom fertilized earlier under current management systems.

Although insemination of cows should begin about 50 to 75 days after calving to maintain a 12-month calving interval, traditional twice-daily observations are likely to fail in identifying about 30% of those ovulating during that period. Mylrea (1962) checked cows in six "well-managed" Australian dairy herds by weekly palpations until they were fertilized. His examinations began 5 weeks after each female had given birth and indicated that quiet ovulations (i.e., with heats not being detected by herdsmen) had occurred in 47, 33, and 25% of cows in three subsequent cycles. Ovulations without estrus being reported declined to about 20% in following cycles.

Hurnik et al. (1975) determined, by hormone assay, that 57 of 69 cows were ovulating within 60 days of calving (high progesterone levels indicated functioning corpora lutea). By using a time-lapse video recorder they were able to observe the behavior of half of their cows for each 24-hour period during 1 hour of viewing time. This approach showed that every ovulation occurring 50 to 75 days after calving was indicated by mounting activity among cows; there were, in fact, no silent ovulations. The other half of the cows were observed twice a day by herdsmen as part of the normal routine; among those animals 35% had ovulations that were undetected by their being mounted.

It appears that truly silent ovulations during the optimal period for rebreeding are uncommon, but that estrus behavior may occur at such times as to go unnoticed in many animals. Hurnik and his colleagues, in analyzing when activity occurred, found that two-thirds of all mounts were so late in the evening or early in the morning that herdsmen would not see them. They recommended additional periods of observation late in the evening and early in the morning, in the absence of feeding and milking distractions.

An alternative procedure is to use testosterone-treated cows as bull substitutes (Signoret, 1975; Kiser et al., 1977). Kiser et al. fitted their treated cows with marking devices and found that during a 20-day experiment 74% of all nonpregnant cows were mounted, indicating estrus, as compared to only 48% observed as being in heat by the herdsman. None of 26 pregnant cows were marked, as expected. Signoret (1975) cautioned that the location of ink marks was important for reliable identification of estrous cows; marks on the rump only were often seen on cows not in heat; they represented unsuccessful mounting attempts. On the other hand, marks on cows' backs in front of the hips indicated that those animals had stood for mounting. The latter conclusion was confirmed by direct observations. Kiser et al., on the basis of several experiments, concluded that "testosterone-treated cows were as effective as surgically altered bulls for detection of estrus."

Genital Stimulation of Estrus

Evidence is cited by Fraser (1968: 60–62) indicating that genital stimulation can be effective in causing anestrous females of a number of ungulate species to show estrous behavior. Such stimulation would ordinarily be provided by the male of the species, but apparently man can substitute in some cases. Thus, 31 of 32 cows with weak or no signs of estrus came into heat in an average of 8 days following massage of the clitoris for up to 7 minutes a day over a 6-day period (Hintnaus, 1965). Twenty of those were inseminated and 11 conceived.

Anestrous cows often show estrus a few days after palpation of the reproductive organs, and irrigation of the uterus causes release of gonadotropic hormones. With those observations in mind, Hays and Calavero (1959) stimulated anestrous cows with an electroejaculation device inserted into the reproductive tract. By such treatment they were able to induce estrus in 40% more cows than was observed in an untreated control group.

PATTERNS OF RECEPTIVITY

Poultry (Chickens and Turkeys)

Chicken and turkey hens do not have estrous cycles as mammalian females do, but their reproductive behavior is set in motion by the same gonadotropic and gonadal hormones (Nalbandov, 1974). Laying hens (chickens) are easily identified as to their hormonal state by their large red combs and all fowl by examination of the cloacal region (moist, soft) and pubic bones (spread apart and flexible) when closely approaching or in the laying phase. As with mammalian species, sexual receptivity is apparent when females approach males or do not move away and assume postures so that mating may occur.

Sexual receptivity in laying but nonmated hens is often apparent to the poultry producer, who observes certain individuals crouching at his approach. It seems likely that those hens have been sexually imprinted to human beings because of their presence during a sensitive period. When roosters are continuously present, crouching is seen only infrequently, presumably because vigorous cocks are frequently testing hens for sexual receptivity by courting behavior, including rear approaches, and under those circumstances crouches merge smoothly into combined crouching–mounting interactions.

Frequency of crouching varies considerably among hens, but Wood-Gush (1958) found no close association between that and rate of egg production. Calculations based on observed mating frequencies in multiple-male flocks of White Leghorns (Kratzer and Craig, 1980) indicate that hens may average at least one mating per day. Such a frequency appears to provide a safety margin, as artificial insemination at weekly intervals maintains essentially the same level of fertilization as does natural mating.

In contrast to chickens, which may complete the mating sequence in a few seconds, turkeys move through their mating behavior in a ponderous fashion. Turkey hens show a marked sexual crouch and in contrast to hens become refractory to additional matings for periods of some days following a mating or after eversion of the oviduct, even if a clumsy male fails to deposit semen. Hale (1955) found the average refractory period to be over 5 days in length, and some individuals did not mate again for periods of 30 days. When the refractory period exceeded 7 days, fertility decreased. It was also observed that female turkeys would occasionally mount and tread in male fashion to the point that "orgasm" was experienced by the mounted hen; refractory periods followed such homosexual interactions in the same manner as when clumsy males failed to inseminate.

Domesticated Mammals

Table 18-3 presents estimates of average age at first estrus, mating patterns, and length of estrous cycles, together with ranges of estrus duration and some behavioral indicators of estrus which are evident to the human observer in the absence of sexually active males. Such a table must be an oversimplification because of the considerable individual variation existing within breeds and large breed differences.

Geographic Effects. Breed differences for seasonally breeding animals are particularly likely to be evident between those kept at different latitudes. Thus, females of breeds developed at greater distances from the equator tend to have shorter seasonal periods of estrous cycling and mating activity is likely to be intense during those periods. Even cattle, generally considered as all-year breeders, may have higher incidences of nonreceptivity or tend to show weaker signs and have reduced duration of estrus during the winter months as the polar regions are approached. Such pronounced seasonality of breeding at high latitudes must have been selected for, as the fitness of animals delivering young during extremely inclement weather would be reduced. The long breeding season of Finnish Landrace sheep (Finnsheep) is a fascinating exception to the general rule.

Length of Estrous Cycle. The interval between comparable phases of different estrous cycles (e.g., beginning of estrus) is remarkably consistent among polyestrous species, at about 21 days (16½ days for sheep). Dogs, unlike most other domestic mammals, tend to have only one rather long estrous period followed by a long anestrous interval, so that only two receptive periods are likely to occur annually and those do not appear to be closely synchronized with the seasons (Fox and Bekoff, 1975). The Basenji breed of dogs is unusual in having only a single estrous period annually, which tends to occur in the early fall.

Duration and Intensity of Estrus. The considerable individual variation in duration of estrus within breeds and species is indicated by the ranges given in Table 18-3. In general terms, virgin females are likely to have shorter heat periods and are relatively easy to displace in advances toward a male if older females are simultaneously in estrus and competing for his attention. Silent or quiet ovulations are not uncommon at the beginning of cycling for virgin females and for older females of seasonally breeding species at the beginning of the season.

Table 18-3 Occurrence, duration, and overt behavioral indicators of estrus in absence of sexually active males[a]

Item	Cattle	Swine	Sheep	Goat	Horse	Cat	Dog
Age at first estrus (average)	9 months	7 months	9 months	5 months	24 months	6 months	10 months
Mating season and cyclicity	All-year polyestrous		Seasonally polyestrous (fall)	Seasonally polyestrous	Seasonally polyestrous (spring and summer)	Seasonally polyestrous (late winter and spring)	Monestrous
Length of estrous cycle (average)	21 days	21 days	16½ days	21 days	21 days	21 days	Twice a year
Duration of estrus (range)	¼–1¼ days	1–3 days	½–2 days	1–4 days	2–10 days	6–10 days	7–9 days
Behavioral indicators							
General activity level	2–4× increase	2× increase	No change	Increased	Increased	Increased	Increased
Vocalization	Increased bellowing	Estrus "grunts"	Increased bleating	Increased bleating	—	Frequent	—
Other females' behavior	Estrous female mounted frequently	Estrous female mounted occasionally	Little interest	Estrous female mounted	None	—	—
Age effects	Estrous periods are generally shorter and behavioral indicators less intense in virgin females; ovulations with "silent" or "quiet" heats may occur frequently in young females in some breeds and species						
Primary criterion in absence of male	Mounting by other cows	Mating stance to pressure on back	None	Rapid wagging of up-turned tail	Clitoris exposed, tail raised, urinating stance	Lordosis, treading, etc., to pressure on back	—

[a]Under moderate to high levels of nutrition in temperate climates; a few breeds may differ from these general descriptions.

REFERENCES

BEACH, F. A., 1976. Sexual attractivity, proceptivity, and receptivity in female mammals. Horm. Behav. 7:105–138.

BHAGWAT, A. L., and J. V. CRAIG, 1979. Effects of male presence on agonistic behavior and productivity of White Leghorn hens. Appl. Anim. Ethol. 5: 267–282.

CHENG, K. M., R. N. SHOFFNER, R. E. PHILLIPS, and F. B. LEE, 1979. Mate preference in wild and domesticated (game farm) Mallards. II. Pairing success. Anim. Behav. 27:417–425.

COLBY, D. R., and J. G. VANDENBERGH, 1974. Regulatory effects of urinary pheromones on puberty in the mouse. Biol. Reprod. 11:268–279.

FLETCHER, I. C., and D. R. LINDSAY, 1971. Effect of rams on the duration of oestrus behaviour in ewes. J. Reprod. Fertil. 25:253–259.

FOX, M. W., and M. BEKOFF, 1975. The behaviour of dogs. In E. S. E. Hafez (Ed.), The Behaviour of Domestic Animals, 3rd ed. Baillière, Tindall Publishers, London.

FRASER, A. F., 1968. Reproductive Behaviour in Ungulates. Academic Press Ltd., London.

HAFEZ, E. S. E., and M. F. BOUISSOU, 1975. The behaviour of cattle. In E. S. E. Hafez (Ed.), The Behaviour of Domestic Animals, 3rd ed. Baillière, Tindall Publishers, London.

HAFEZ, E. S. E., M. W. SCHEIN, and R. EWBANK, 1969. The behaviour of cattle. In E. S. E. Hafez (Ed.), The Behaviour of Domestic Animals, 2nd ed. Baillière, Tindall & Cassell Ltd., London.

HAFEZ, E. S. E., L. J. SUMPTION, and J. S. JAKWAY, 1962. The behaviour of swine. In E. S. E. Hafez (Ed.), The Behaviour of Domestic Animals. Baillière, Tindall & Cox Ltd., London.

HALE, E. B., 1955. Defects in sexual behavior as factors affecting fertility in turkeys. Poult. Sci. 34:1059–1067.

——, W. M. SCHLEIDT, and M. W. SCHEIN, 1969. The behaviour of turkeys. In E. S. E. Hafez (Ed.), The Behaviour of Domestic Animals, 2nd ed. Baillière, Tindall & Cassell Ltd., London.

HART, G. W., S. W. MEAD, and W. M. REGAN, 1946. Stimulating the sex drive of bovine males in artificial insemination. Endocrinology 39:221–228.

HAYS, R. L., and C. H. CALAVERO, 1959. Induction of estrus by electrical stimulation. Am. J. Physiol. 196:899–900.

HILLYER, G. M., 1976. An investigation using a synthetic porcine pheromone and the effect on days from weaning to conception. Vet. Rec. 98:93–94.

HINDE, R. A., 1965. Interaction of internal and external factors in integration of canary reproduction. *In* F. A. Beach (Ed.), Sex and Behavior. John Wiley & Sons, Inc., New York.

HINTNAUS, J., 1965. Reflex-induced oestrus in cattle with weak or no signs of heat. Vet. Med. Prague 10:69–76. (Anim. Breed. Abstr., 1965, 33:2290.)

HURNIK, J. F., G. J. KING, and H. A. ROBERTSON, 1975. Estrous and related behaviour in postpartum Holstein cows. Appl. Anim. Ethol. 2:55–68.

KIDDY, C. A., D. S. MITCHELL, D. J. BOLT, and H. W. HAWK, 1978. Detection of estrus-related odors in cows by trained dogs. Biol. Reprod. 19:389–395.

KIRACOFE, G. H., 1980. Uterine involution: its role in regulating post-partum intervals. J. Anim. Sci., Suppl.: XIV Bienniel Symp. on Anim. Reprod. (in press).

KISER, T. E., J. H. BRITT, and H. D. RITCHIE, 1977. Testosterone treatment of cows for use in detection of estrus. J. Anim. Sci. 44:1030–1035.

KRATZER, D. D., and J. V. CRAIG, 1980. Mating behavior of cockerels: effects of social status, group size and group density. Appl. Anim. Ethol. 6:49–62.

LAUDERDALE, J. W., B. E. SEGUIN, J. N. STELLFLUG, J. R. CHENAULT, W. W. THATCHER, C. K. VINCENT, and A. F. LOYANCANO, 1974. Fertility of cattle following $PGF_{2\alpha}$ injection. J. Anim. Sci. 38:964–967.

LeBOEUF, B. J., 1967. Interindividual associations in dogs. Behaviour 29:268–295.

LEHRMAN, D. S., 1965. Interaction between internal and external environments in the regularity of the reproductive cycle of the ring dove. *In* F. A. Beach (Ed.), Sex and Behavior. John Wiley & Sons, Inc., New York.

LILL, A., 1968a. Some observations on the isolating potential of aggressive behaviour in the domestic fowl. Behaviour 31:127–143.

——, 1968b. An analysis of sexual isolation in the domestic fowl: II. The basis of homogamy in females. Behaviour 30:127–145.

——, and D. G. M. WOOD-GUSH, 1965. Potential ethological isolating mechanisms and assortative mating in the domestic fowl. Behaviour 25:16–44.

LINDSAY, D. R., 1965. The importance of olfactory stimuli in the mating behaviour of the ram. Anim. Behav. 13:75–78.

——, and I. C. FLETCHER, 1972. Ram-seeking activity associated with oestrous behaviour in ewes. Anim. Behav. 20:452–456.

LINDSAY, D. R., and T. J. ROBINSON, 1961. Studies on the efficiency of mating in the sheep. II. The effect of freedom of rams, paddock size, and age of ewes. J. Agric. Sci. 57:141–145.

MATTNER, P. E., J. M. GEORGE, and A. W. H. BRADEN, 1974. Herd mating activity in cattle. J. Reprod. Fert. 36:454-455.

MAVROGENIS, A. P., and O. W. ROBISON, 1976. Factors affecting puberty in swine. J. Anim. Sci. 42:1251-1255.

MYLREA, P. J., 1962. Clinical observations on reproduction in dairy cows. Aust. Vet. J. 38:253-258.

NALBANDOV, A. V., 1974. Egg laying. *In* H. H. Cole and M. Ronning (Eds.), Animal Agriculture. W. H. Freeman and Company, Publishers, San Francisco.

PARSON, S. D., and G. L. HUNTER, 1967. Effect of the ram on the duration of oestrus in the ewe. J. Reprod. Fertil. 14:61-70.

PEPELKO, W. E., and M. T. CLEGG, 1965. Studies of mating behaviour and some factors influencing the sexual response in the male sheep *Ovis aries*. Anim. Behav. 13:249-258.

SCHEIN, M. W., and M. H. FOHRMAN, 1955. Social dominance relationships in a herd of dairy cattle. Br. J. Anim. Behav. 3:45-55.

SCHINCKEL, P. G., 1954. The effect of the ram on the incidence and occurrence of oestrus in ewes. Aust. Vet. J. 30:189-195.

SIGNORET, J. P., 1970a. Reproductive behaviour in pigs. J. Reprod. Fertil., Suppl. 11:105-117.

——, 1970b. Sexual behaviour patterns in female domestic pigs (*Sus scrofa* L.) reared in isolation from males. Anim. Behav. 18:165-168.

——, 1971. The reproductive behaviour of pigs in relation to fertility. Vet. Rec. 88:34-38.

——, 1975. Nouvelle méthode de détection de l'estrus chez les bovins. Ann. Zootech. 24:125-127.

——, B. A. BALDWIN, D. FRASER, and E. S. E. HAFEZ, 1975. The behaviour of swine. *In* E. S. E. Hafez (Ed.), The Behaviour of Domestic Animals, 3rd ed. Baillière, Tindall Publishers, London.

SIGNORET, J. P., F. Du MESNIL Du BUISSON, and P. MAULEON, 1972. Effect of mating on the onset and duration of ovulation in the sow. J. Reprod. Fert. 31:327-330.

VANDENBERGH, J. G., 1967. Effect of the presence of a male on the sexual maturation of female mice. Endocrinology 81:345-349.

WHALEN, R. E., 1963. Sexual behavior of cats. Behaviour 20:321-342.

WOOD-GUSH, D. G. M., 1958. Fecundity and sexual receptivity in the Brown Leghorn female. Poult. Sci. 37:30-33.

19

Parturition, Suckling, and Weaning

PARTURITION: UNDERLYING MECHANISMS

Synchrony and Timing of Births

Seasonal breeding is the common mode of reproduction in natural environments. An obvious advantage of births occurring within a short period is that optimal nutritional conditions for omnivorous and herbivorous species do not last for long in most geographic areas, and such conditions are important to the survival and well-being of the young. There is another advantage in having the young of different females born at about the same time. The presence of a "glut" of vulnerable, newborn animals is an obvious evolutionary strategy; if more are born within a short period than predators can destroy, there is a better chance of survival than if births are spread evenly over a longer period.

As indicated earlier, when one or more males join with a band of females, particularly among sheep, synchrony of estrous cycles occurs. Therefore, the social environment, as well as seasonal priming (usually associated with day length), indirectly influences when parturition will occur, if birth follows fertilization after a fixed interval.

Diurnal activities or disturbances caused by caretakers may influence the portion of the day when more young are born, especially in horses. Of 501 Thoroughbred mares whose hour of foaling was known, 86% gave birth during the mostly dark hours of 7 P.M. to 7 A.M. (Rossdale and Short, 1967). For housed cattle, 66% were reported to have calved during the 12-hour period centering on midnight (Arthur, 1961, in Arnold and Dudzinski, 1978). For Hereford cows kept in an outside lot, 31 of 38 (82%) calved during the 12-hour period beginning at 2 P.M. and continuing into the early hours of the morning (George and Barger, 1974). The early hours of darkness appear to be the most likely time for farrowing in pigs (Alexander et al., 1974). The situation in sheep is not at all clear; but frequency of lambing during the dark hours does not appear to be greater than during the daytime. Among sheep, breed and husbandry conditions may have importance in whether "peaks" in lambing time occur during the 24-hour daily cycle (Arnold and Morgan, 1975).

General Consideration: Time of Day and Parturition. The overall picture for horses, cattle, and swine appears to be that more young are born during the hours of darkness, but some species are more affected than others. Fraser (1968) postulated that adrenaline, produced by aroused animals, may be responsible for inhibiting birth during periods of greater activity or excitement. During the quieter hours adrenaline levels fall, and Fraser suggests that the action of the hormone oxytocin (associated with uterine contractions needed for expulsion of the fetus) would no longer be blocked by adrenaline, so that the birth process could then be completed.

Internal Environment

Changes during Gestation. Pregnancy occupies a considerable time span in the life of most domesticated mammalian females, but gestation periods are obviously of shorter duration for those giving multiple births. Although hormonal changes of considerable magnitude occur, they do so gradually, until just before parturition (Catchpole, 1969). Other than the obvious absence of cyclic estrous periods and, in some species, increased attraction to the newborn of other females shortly before parturition, the pregnant female does not differ greatly in behavior from the nonpregnant female.

The fetal portion of the placenta and endocrine glands of the fetus are of major importance in sustaining pregnancy and in the initiation of parturition. Fetal genotype influences gestation length, as shown by comparisons of horses, asses, and their crosses (Asdell, 1964; Short, 1960). Thus, mares, which usually give birth at about 340 days, average 15 additional days before giving birth to a mule foal, and jennets foal at about 350 days if carrying a hinney as compared to 365 days for a donkey foal.

It has been shown that destruction of the fetal lamb's pituitary gland results in continuing pregnancy beyond the normal time of parturition (Liggins et al., 1967). Later studies indicated that the fetus' own hypothalamic-pituitary system stimulated its adrenal cortex to secrete glucocorticoids, which in some way are significant in triggering the onset of parturition by the mother (Liggins, 1968). Similar mechanisms appear to be involved in the goat (Thorburn et al., 1972) and the sow (Dvorak, 1972). Although changes in levels of other hormones (progesterone, estrogen, prostaglandin $F_{2\alpha}$, etc.) are involved in various ways in preparing the expectant mother, it appears that termination of pregnancy is largely controlled by an increase in fetal corticoids (Liggins et al., 1972).

The hormone oxytocin, known to be involved in the final stages of the birth process, appears to be secreted abruptly by the posterior pituitary in response to stimuli provided by the fetus as it enters the lower part of the genital tract (Chard, 1972).

Summary

Timing mechanisms for the onset of reproductivity, duration of pregnancy, and parturition operate at several distinct levels for seasonally breeding ungulates. Some of the same influences are important in other species, also. In terms of seasonal breeding activity, day length appears to be of major importance, at least in the temperate zones (Chapter 8); the same can be said for birds. The social environment exerts an important, but secondary, influence; thus, when adult males leave their largely bachelor life and join female herds of swine, sheep, and cattle on a seasonal basis, there is an acceleration of onset and synchrony of female reproductive cycling which would not occur in their absence (Chapter 18). With pregnancy established, gestation proceeds according to each species' own timetable, so that the young will be delivered in a time of nutritional plenty. Hormonal changes associated with the termination of pregnancy are primarily of fetal origin and the fetus' own genotype plays a role. The final onset of labor is clearly associated with time of day in horses, there is a lesser effect in swine and cattle, and sheep appear to be largely or entirely unaffected in this way.

PARTURITION

Ability on the part of the caretaker to recognize behavioral "signs" that birth is about to occur can be valuable in providing for the needs of the mother and her offspring. When the dam is confined with a group and unable to seek suitable shelter for herself, it becomes especially important. Stronger mother–young bonding, less opportunity for interference and possible "stealing" of newborn animals by other females about to give birth (especially among ewes), and better care by the dam are benefits associated with temporary separation. Greater protection from the elements during severe weather also increases chances of survival if temporary shelter during and shortly after birth can be provided. Arnold and Morgan (1975), studying causes of lamb mortality in Western Australia, found that 18% of all lambs born died, and 52% of those born on rainy days did not survive.

Prepartum Behavior

Apart from seeking varying degrees of seclusion and the building of nests by litter-bearing animals, if materials are available (Table 4–2), restlessness usually becomes evident for a few hours before the young are born (Table 19-1). Increasingly frequent changes of position were seen in sows approaching parturition until they were moving about every few minutes (Jones, 1966). Pain increases as the cervix, which closes off the uterus from the vagina during pregnancy, becomes distended during the first stage of labor and muscular contractions begin. Fraser (1968) described how mares frequently look back toward their flanks and cows appear apprehensive, looking all about, possibly kicking at the abdomen, and treading with the hind feet. Ewes are less likely to give overt indications that the first stage of labor is in progress; Arnold and Morgan (1975) found behavioral signs of imminence of parturition in only 64% of carefully observed ewes.

Birth

Expulsion of the young occurs during the second stage of labor (i.e., after dilation of the cervix). Pain is most evident during this stage, as muscular contractions of the abdomen and uterus force the fetus through the cervix, into the vagina, and then out through the vulva. As the fetus leaves the uterus, the chorionic membrane, which has nourished the fetus, remains attached and ruptures as that fluid escapes, and as stockmen say, "the first water-bag bursts." The amniotic bladder, together with the fetus, then moves through the remainder of the birth canal, as repeated periods of straining, interrupted by resting periods, force them out. The amniotic bladder typically appears through the vulva and often breaks before the fetus appears. The forefeet and head normally appear first in most

Table 19-1 Typical parturient behavior of some domesticated animals

Item	Species					
	Cow	Ewe	Mare	Sow	Bitch	Cat
First stage of labor	Separate from group if possible. Restless, become attached to site				Finish "nest building" if material is available	
Dilation of cervix				Defends "nest"	Anxiety (may want human company)	
Second stage						
First water bag bursts	Interest in birth fluids—lick, smell ("flehmen" response)					
	Pain evident—turning head to rear, groans, muscular contractions					
Delivery	Recumbent. After hips pass vulva, may stand; remainder of fetus emerges easily			Recumbent—lateral to ventral, etc.		
	Stands after delivery	Stands within 1 min of delivery	Rests 5–15 min before rising	Rises after last of litter is born		
Total delivery time (average)	1–4 hr	½–1 hr	¼–½ hr	2–3 hr	2–3 hr	2–3 hr
Third stage Delivery of placenta after giving birth	Several hours	Up to 5 hr	1 hr		Several hours	
Placenta eaten?	Yes (usually)	No (may eat membranes adhering to young)	No	Yes; may eat dead fetuses	Yes	Yes

334

species, but anterior and posterior presentations occur with nearly equal frequency in swine (Randall, 1972). Emergence of the head completes the most difficult period of birth and the remainder of the fetus often slips out with relative ease. The relative difficulty of birth and duration of the delivery process varies among species; litter-bearing animals require more time (Table 19-1).

Postpartum Behavior

The third stage of labor involves expulsion of the fetal membranes or afterbirth and usually occurs without much apparent effort, several hours after the young arrive. An interesting phenomenon is that dams of young that are kept at the site of birth for several days following birth typically lick up or consume any bedding contaminated by birth fluids and also eat the placenta. On the other hand, ewes and mares, which lead their young away soon after birth, do not consume the placenta. Because the placenta appears to offer no real nutritional value, particularly to ruminants such as the cow, Fraser (1968) deduced that placental consumption in such species must be a behavioral adaptation that reduces the likelihood of predators being attracted to the site of birth by the odor of decomposing afterbirth. It also seems likely that keeping the nest or birth site cleared of waste material may protect the young from contamination and infection by microorganisms.

INITIAL GROOMING AND SUCKLING

After a brief period of rest, new mothers show much concern for their newborn by vigorously licking and cleaning them of birth fluids. Licking appears helpful in establishing vigorous breathing, in reducing chances of chilling, and in initiating eliminative behavior. Following the initial grooming by the mother, the young actively seek out the udder and ingest antibody-rich colostrum (Table 19-2). As the young animal suckles, both initially and later, the dam turns her head and sniffs her progeny (Figure 9-2). As indicated in Chapter 9, the initial cleaning of the young allows the dam to identify her own progeny and determines whether a young animal attempting to suckle will be accepted.

Suckling typically begins in ungulates soon after the mother has completed grooming her offspring and the young animal has struggled to its feet and successfully located a nipple. Puppies and kittens crawl to the udder. Because of the need for grooming and the young

Table 19-2 Some aspects of maternal care and suckling behavior in cattle, sheep, horses, and swine*

Item	Cattle — Beef Cow	Cattle — Dairy Cow	Cattle — Dairy Heifer	Sheep (single lamb)	Horse	Swine
Initial grooming (hr)	¾[a]	½	¼	¼	½[b]	0
Adopt young of others?		Unlikely after grooming own progeny				Yes, 2+ days
Interval, following birth (hr)						
Newborn stands	½[c]	1	1¼	¼[d]	1[e]	2 min[f]
Newborn suckles	1¼	4¼[d]	3½	1	2	¼ hr
Suckling frequency						
First few days		5-8×/24 hr[g]		15×/12 hr[h]	4×/hr[b]	1×/hr[i]
Midlactation		3-6×/24 hr[d]		10×/12 hr	1×/hr	1×/hr
Suckling duration/bout						*Initial "nosing"* / *Suckling*
First few days		8-10 min[j]		40 sec[h]	70 sec[b]	60 sec[i] / 25 sec
Midlactation		8-10 min		15 sec	60 sec	90 sec / 20 sec
Age at natural weaning		10 months[k]		4-6 months[l]	11+ months[b] (pregnant mares)	?

*Numerical values are rounded averages. Where breeds vary considerably or large differences occur between dams of different ages, a range of mean values may be given.

Note: Superscripts indicate code letters for references, as follows: [a] Selman et al. (1970a); [b] Tyler (1972); [c] Selman et al. (1970b); [d] Arnold and Morgan (1975); [e] Rossdale (1967); [f] Randall (1972); [g] Walker (1950); [h] Ewbank (1967); [i] Barber et al. (1955); [j] Hafez and Lineweaver (1968); [k] Schloeth (1961); [l] Geist (1971). Where all figures within a cell of the table are derived from the same reference, the code letter is indicated only once.

animal's exploratory efforts in finding a nipple, the first milk may not be ingested until 1 to several hours after birth.

Sow-Piglet Interactions

Swine differ considerably from other domesticated mammals in the mother's early response to the young. Piglets are not cleaned by their dam at birth. The newborn pig is very much on its own. Most piglets stand within 2 minutes of delivery and, sucking at every protuberance on the sow, move about exploring her ventral surface until a teat is located and colostrum is ingested. It is believed that a sow learns to identify her own pigs primarily by the sense of smell, but pigs may be moved from sow to sow for at least 2 days following farrowing without being rejected. Although sows seem to form bonds to specific piglets slowly, they soon respond to alarm cries and defend their litters.

Randall (1972) noted that several first-litter gilts became mildly disturbed after a few pigs were delivered, and this was usually caused by a newborn pig approaching the mother's head for the first time. Those gilts responded as if frightened; they would back away from the piglet, barking, and would sometimes bite at it. In 3 of 60 gilts observed, even more aggressive behavior was seen; those dams attacked piglets that were moving about the pen. This aggression stopped as soon as one or more piglets found a teat and began suckling. Although piglet killing and cannibalism are not common, it does occur, usually in less than half of 1% of first litters (R. Hines, personal communication). Temporary removal of the young usually calms the gilt, so that they can then be safely returned.

SUCKLING AND WEANING

Some characteristics of suckling behavior and approximate ages at natural weaning are indicated in Table 19-2. Prior to weaning, young animals must learn to ingest solid foods if the transition is to be gradual and not associated with a nutritional setback when separated from their dams. How rapidly this will occur is associated with availability of other foods, relative developmental stage of the species at birth, and subsequent rate of development. Lambs and foals may be nibbling at herbage within a few days, but puppies and kittens are dependent on mothers' milk for the first couple of weeks. However, feeding of solid food to puppies and kittens begins soon after they develop sight and control of their movements; dams that are predators bring food home for their young.

Cattle

With reference to care of the young, cattle have been called an "outlying" or "hider" species, in contrast to "follower" species such as sheep and horses. With outlying species the young are typically left alone for periods of several hours while the dam forages. Suckling behavior of cattle reflects the mother–young life-style; calves suckle only five to eight times daily for the first few days, but have relatively long periods of suckling when it does occur. Suckling frequency decreases with increasing age, but not dramatically.

Artificial selection of dairy cattle appears to have produced marked changes in maternal care and in the ability of calves to obtain milk initially. Selman et al. (1970a, 1970b) have compared the behavior of cows and calves of beef breeds with that of cows and heifers and their calves of a dairy breed (Ayrshire) during the first 8 hours after birth (Table 19-2). Most dams were observed to begin licking their calves immediately after delivery, but the duration of initial, continuous licking was significantly less for dairy than for beef cows. Dairy heifers licked their calves for less than half as long as dairy cows. Fetal waste (meconium), passed by the calf during initial suckling, was carefully licked up by all dams. Because of the licking activity, calves were usually at the head of the dam when first struggling to their feet.

The shape of a cow's underbelly appeared to Selman et al. (1970a) to be important in determining how long it would take a vigorous calf to locate the udder and begin suckling. Therefore, they classified pregnant females in their study as having good or poor body shape with reference to supposed ease of finding the udder by the calf. "Good shape" classification required that the udder and teats were either level with or higher than the posterior portion of the cow's breastbone (xiphisternum); cows with "poor shape" had large abdomens or udders and teats extending below the line of the breastbone. As expected, in exploring their dams' bodies, newborn calves showed a marked tendency to thrust their muzzles under the cow's abdomen and then push their noses as high as possible. Thus, calves usually spend considerable time exploring the highest parts of their dams' underbellies. For cows with "poor shape" [i.e., having large abdomens and udders (more typical of dairy cows than of beef dams)], calves tended to focus their exploratory activity around the forelegs or high above the teats. However, calves exploring dams with "good shape" quickly located the udder in most cases. Therefore, differences in body shape between beef and dairy cows appear to be largely responsible for the very large difference in time required before first suckling from beef and dairy cows (1¼ hours for beef vs. 4¼ hours for dairy). Five of eight calves failing to suckle within the first 8 hours after birth did so because of poor dam shape.

Early Separation of Mother and Young. The importance of suckling from the dam initially, rather than being fed colostrum from a bucket, has been identified as a significant influence in mortality of dairy heifer calves. Thus, Withers showed in a series of reports (e.g., Withers, 1953) that mortality of calves was 3.9% when obtaining colostrum directly from their dams as compared to 9.1% when fed colostrum from a bucket. Selman et al. (1971) compared calves fed colostrum from teat buckets in the presence or absence of their dams. Those calves kept with their dams for at least 18 hours, but muzzled to prevent natural suckling, had significantly greater absorption of immune lactoglobulin. Thus, the psychological environment of the young calf is important in affecting resistance to certain diseases because of the greater or lesser absorption of protective antibody from colostrum. If newborn dairy calves are not observed closely, it may not be known whether they have in fact nursed during the 6 to 12 hours that many dairymen allow before separating them from their dams. Selman et al. (1970b) found that of 30 calves observed continuously for the first 8 hours after birth, nearly 25% had not suckled.

When the dairy calf is removed from its dam shortly after birth, the dairyman takes over the role of "mother." Owen (1978) emphasizes the need to assure the calf of an adequate intake of colostrum. Because the dairy cow has been successfully bred for a large digestive system and large udder, many calves may not have successfully located the udder or consumed enough colostrum by the time of separation. Owen suggests, therefore, that the calf be fed known minimal amounts of colostrum (4 to 6 lb or about 2 to 3 kg) twice at 12-hour intervals if removed from the cow within 6 to 12 hours after birth.

Early removal of a newborn mammal from its mother always seems to cause abnormal behavior related to suckling activity. When unable to suck on a teat or suitable substitute, young mammals will suck on the navel, scrotum, ear, or tail of other animals when reared in a group. Because such activity may cause infections of the parts sucked and because the ingestion of foreign material may be harmful, very early weaned animals are likely to be kept in individual stalls or cages. Calves exhibit less emotional upset when reared in separate compartments than do most other mammals, presumably because it is more natural; very young calves typically remain quietly by themselves for hours at a time while their dams are absent, as already noted.

Should dairy calves be removed from their dams immediately after birth and cared for solely by man? Some believe that such a procedure reduces the emotional upset of the cow as compared to removing the calf after the mother has bonded to it through several hours of close association. However, as indicated above, calves live better if allowed to obtain their first colostrum from their mother rather than from a bucket. Another aspect of this situation was shown by Fraser (1974), who found that the retention of fetal membranes by cows not allowed to suckle their newborn calves was 2 to 3% higher than where suckling by the calves was allowed.

Time of separation of dairy heifers from their dams and methods of rearing in isolation or in groups on subsequent behavior and milk production have been studied at Purdue University. In one study (Donaldson et al., 1972), an attempt was made to remove all calves from their dams within 30 minutes of birth and before being nursed by their dams. Calves were then subdivided into four treatment groups, so that all four combinations of being isolated or grouped during feeding and nonfeeding periods were represented. All calves were later placed together, from 18 weeks until parturition. Behavior of the heifers toward their calves was strikingly different, depending on their rearing experience; those reared in isolation (except for considerable human contact) cleaned their calves better, allowed more nursing within 4 hours of birth, were more protective, and did not vocalize shortly after birth. Heifers reared together were less motherly in every measurement taken. Differences between treatment groups were less pronounced with second calves. Milking-parlor behavior did not differ between treatment groups, but large differences were found in milk production. Although Holstein cows fed and kept in isolation from birth until 18 weeks of age produced more milk than Holsteins of the three other treatment groups, such a relationship was not found among Holstein × Red Danish crossbreds. Obviously, the dramatic effects of early experience on maternal behavior and milk production obtained in this study deserve further experimental verification and examination of factors responsible.

A second study by the Purdue group involved calves removed immediately at birth (in which the herdsman served as a surrogate mother by cleaning and feeding the newborn calf), a second group of calves that were left with their dams for 24 hours, and a third group removed at 72 hours postpartum (Albright et al., 1975, and personal communication). Those heifer calves removed at birth tended to stand sooner after giving birth to their own calves, began licking sooner, and licked their calves longer. (However, those differences were not significant.) Milking temperament was superior for those separated from their dams immediately after birth, but those left with their dams for the first 3 days had highest milk yields. Apparently, those calves given intimate care by man soon after birth formed a close and long-lasting social bond to man, but those remaining with their own dam for 3 days may have received some benefit from maternal care of their own species. As in the other Purdue study, these results indicate intriguing avenues for further investigations.

The importance of intimate human contact very early in the dairy heifer's life in terms of future ease of handling and dairy temperament have been indicated. Nevertheless, technological changes may decrease, rather than increase, human–dairy heifer contacts. For example, several reports (e.g., Appleman and Owen, 1975) have indicated that once-daily feeding may be as successful as twice-daily feeding. Automatic feeding devices for calves may further reduce human contact during their early life. Indeed, it may be fortunate, from the standpoint of future milking temperament, that the very young calf removed from its mother and fed from a bucket must be taught to drink. This necessity usually requires close contact of the caretaker and the infant heifer for at least the first few days.

Sheep

In marked contrast to cattle, ewes and their lambs maintain close contact for extended periods. Even during the first day of life, the lamb may follow its dam for a long distance. Close observation of mother–young groupings (Morgan and Arnold, 1974) indicated that when a ewe was lying, the lamb was also lying nearby on 93% of all such occasions; this relationship did not change with age and the distance separating them was usually less than 1 meter.

Suckling frequency for lambs during the first week of life is more frequent and lasts about three times as long per bout as compared to midlactation (Table 19-2). Apparently, the baby lamb is allowed to suckle at any time and for as long as it desires (Munro, 1956). However, after the first week or two, ewes may restrict the frequency and duration of suckling by walking away. Ewes with twin lambs appear to resist sucklings by single members of the pair after the first week; both must be present before the ewe will stand.

Twins suckle at higher frequencies than do single lambs for the first 4 weeks of life. The cause is not clear; either social facilitation or differences in milk production may be involved. Ewes with twins produce more milk, but not twice as much, so that twin lambs may be more hungry. Ewbank (1967) suggests that the decline in extra sucklings for twins after the fourth week may result from ewes refusing unlimited access to the udder, or twin lambs may have been forced to find other sources of food to satisfy their hunger.

Lambs appear to develop a preference for a particular side of the ewe for suckling, but this develops slowly, not being very evident until about the fifth week (Ewbank, 1967). Although tending to suckle from one side only, single lambs nevertheless appear to use both teats equally. When one of a pair of twins is removed, the remaining member quickly learns to use both sides of the udder (Ewbank and Mason, 1967).

Horses

Among the semiwild New Forest ponies observed by Tyler (1972), the mare's behavior toward her foal on the day of birth was directed to remaining very near to it, especially when other ponies were near. Foals often trotted or galloped short distances with their mothers or even after other mares on the first day. However, when attracted to other mares, they were rushed after by their own dam and separated from them. Adult mares were often aggressive to foals other than their own. By the second day, foals followed only their own mothers.

During the first few suckling bouts, New Forest mares would cooperate with their foals by standing still and making the nipples easily available. Thereafter, mares were likely to limit the frequency or length of suckling bouts by moving away. Following persistent efforts by the foal, a mare would usually stand and allow suckling. Frequency of suckling during daytime observations was four times per hour during the first week, but it decreased steadily to twice an hour by the sixth week, once an hour by the fifth month, and once every 2 hours from 8 months until weaning, a few days or weeks before mothers foaled again (Table 19-2). Tyler believes that the act of suckling must be comforting in itself, whether milk is obtained or not; several foals were observed sucking nipples of older sisters that had not foaled. Mares sometimes allowed 1- or 2-year-olds to continue suckling even after the birth of another foal. Most suckling bouts lasted about a minute.

Swine

Piglets are extremely precocial at birth as compared to other domestic animals, particularly when compared with other multiple-birth species, such as cats and dogs. After struggling to their feet within a minute or two of birth, they explore the body of the sow and usually locate a nipple within 10 to 15 minutes. Baby pigs compete for nipples and those arriving late in the farrowing sequence are at a disadvantage (Chapter 12). Late arriving and "runt" pigs in large litters may fail to gain access to a productive teat or, in some cases, to a teat at all.

Suckling bouts during early and midlactation occur at about hourly intervals (Table 19-2), but intervals tend to lengthen in later stages of lactation to 1½ to 2 hours (Barber et al., 1955). Although piglets may hold nipples in their mouths for long periods, especially early in development, milk letdown is relatively brief (20 to 25 seconds) and occurs only after vigorous "nosing" of the udder (Figure 19-1). The amount of nosing required to stimulate the sow to let her milk down gradually increases from about a minute during the first week to about 2 minutes by the eighth week of lactation. The exception is that milk flows freely during parturition without the nosing stimulus, presumably because of the flow of oxytocin, which has the dual role of causing uterine contractions during parturition and of being involved in the initiation of milk flow.

Figure 19-1 Piglets at the udder may hold the nipple in their mouths for long periods. Milk letdown is relatively brief and occurs only after a period of vigorous "nosing" by the litter of pigs. (Photograph by the author.)

Very Early Separation. High mortality is likely to occur in either very small or very large litters. When three or fewer pigs are present, sows tend to stop lactating, probably because less than the threshold amount of udder stimulation occurs; thus, all pigs died in 25 of 37 such litters studied by Pomeroy (1960). Mortality in large litters may result from lack of enough functional nipples or because of the presence of smaller and weaker pigs, which are at a competitive disadvantage in suckling and also are subject to greater heat loss. Although baby pigs may die from a variety of causes following parturition, it appears that starvation is the most likely (English and Smith, 1975).

Baby pigs are born with essentially no antibodies to protect them, but passive immune protection is provided by ingesting colostrum. Maternal antibodies, having a half-life of about 22 days, are absorbed through the piglet's intestinal epithelium, but such absorption rapidly decreases so that little occurs by 3 days after birth. Nevertheless, continuing access to colostrum, provided from cow's milk, has been found to be highly beneficial to the piglet separated from its dam, by providing "local" protection to the intestinal surface, even though not absorbed (Lecce, 1975).

It appears that piglets allowed to nurse the sow for at least 15 to 24 hours may then be removed to a temperature-controlled room and successfully reared on "practical" diets, including cow's colostrum for the first few days (Noll, 1979). Piglets readily learn to drink from a bowl at 1 day of age if they have previously been deprived of liquid and then have their noses placed in it during the first one or two feedings. Although special care is obviously required soon after birth to ensure that piglets receive adequate sow's colostrum initially, the method appears likely to be adopted increasingly to save piglets that otherwise might not survive (Figure 19-2). Effects of early separation from the sow and physical isolation from littermates on behavior deserve attention. Noll found that such pigs were equal in weight to sow-reared pigs at 3 weeks, and subsequent gains were similar for both types of rearing.

MAN'S ROLE

Domestication of animals, by definition, indicates that man plays a significant role in influencing or controlling the breeding, care, and feeding of his animals. Control of breeding, necessary for selection purposes, was achieved in the past primarily by the castration or elimination of excess males. With species of high reproductive

Figure 19-2 Piglets can be successfully reared in cage batteries (in warm rooms) after obtaining colostrum from their dam and cow's colostrum for the first week to 10 days. (Photograph by the author.)

potential, such as poultry and litter-bearing animals, a significant number of females may also be eliminated in the selection process. Artificial insemination, especially in dairy cattle and turkeys, added greater control and possibilities of selection. In recent decades, methods of synchronizing ovulation and parturition have been developed and are likely to be adopted in some areas of intensive livestock production. In similar fashion, technological innovations allow the use of females as egg-producing (poultry) or milk-producing (dairy cattle and goats) specialists. Because of selection and mechanical devices, the hen has become obsolete as a mother; artificial incubators and brooders have taken her place. The dairy cow has also become deficient in motherly ways and her calf is usually removed within a day or two of birth. The sow, with the help of the dairy cow's excess colostrum, may in the future be destined to produce litters of pigs with similar brief contact with her offspring.

With all these changes, the dependence of the animal upon man increases.

There is evidence indicating that the behavioral needs of animals have sometimes been neglected when natural life-styles are replaced by artificially contrived ones. More attention to and study of animals' social and other behavioral requirements would be mutually beneficial to both man and beast. If those needs can be met more adequately, animals will be easier to handle, stress will be reduced, and performance or productivity improved. Enriched early environments with adequate exposure to man allow primary socialization and conditioning procedures that may be especially helpful where animals are to be kept as companions, work animals, or for productivity in high-density, intensive management systems. Genetic selection may also be a powerful ally in adapting animals to environments unlike those in which the species evolved.

REFERENCES

ALBRIGHT, J. L., C. M. BROWN, D. L. TRAYLOR, and J. C. WILSON, 1975. Effects of early experience upon later maternal behavior and temperament in cows. J. Dairy Sci. 58:749 (Abstract).

ALEXANDER, G., J. P. SIGNORET, and E. S. E. HAFEZ, 1974. Sexual and maternal behavior. *In* E. S. E. Hafez (Ed.), Reproduction in Farm Animals, 3rd ed. Lea & Febiger, Philadelphia.

APPLEMAN, R. D. and F. G. OWEN, 1975. Breeding, housing and feeding management. J. Dairy Sci. 58:447–464.

ARNOLD, G. W., and M. L. DUDZINSKI, 1978. Ethology of Free-ranging Domestic Animals. Elsevier Scientific Publishing Co., Amsterdam.

ARNOLD, G. W., and P. D. MORGAN, 1975. Behaviour of the ewe and lamb at lambing and its relationship to lamb mortality. Appl. Anim. Ethol. 2: 25–46.

ARTHUR, G. H., 1961. Some observations on the behavior of parturient farm animals with particular reference to cattle. Vet. Rev. 12:75–84.

ASDELL, S. A., 1964. Patterns of Mammalian Reproduction. Cornell University Press, Ithaca, N.Y.

BARBER, R. S., R. BRAUDE, and K. G. MITCHELL, 1955. Studies on milk production of Large White pigs. J. Agric. Sci., Camb. 46:97–118.

CATCHPOLE, H. R., 1969. Hormonal mechanisms during pregnancy and parturition. *In* H. H. Cole and P. T. Cupps (Eds.), Reproduction in Domestic Animals, 2nd ed. Academic Press, Inc., New York.

CHARD, T., 1972. The posterior pituitary in human and animal parturition. J. Reprod. Fertil., Suppl. 16:121-138.

DONALDSON, S. L., J. L. ALBRIGHT, and W. C. BLACK, 1972. Primary social relationships and cattle behavior. Proc. Indiana Acad. Sci. 81:345-351.

DVOŘÁK, M., 1972. Adrenocortical function in foetal, neonatal and young pigs. J. Endocrinol. 54:473-481.

ENGLISH, P. R., and W. J. SMITH, 1975. Some causes of death in neonatal piglets. Vet. Annu. 15:95-104.

EWBANK, R., 1967. Nursing and suckling behaviour amongst Clun Forest ewes and lambs. Anim. Behav. 15:251-258.

——, and A. C. MASON, 1967. A note on the suckling behaviour of twin lambs reared as singles. Anim. Prod. 9:417-420.

FRASER, A. F., 1968. Reproductive Behaviour in Ungulates. Academic Press, Inc., New York.

——, 1974. "Ethostasis": a concept of restricted behaviour as a stressor in animal husbandry. Br. Vet. J. 130:91-92.

GEIST, V., 1971. Mountain sheep, a study in behaviour and evolution. University of Chicago Press, Chicago.

GEORGE, J. M., and I. A. BARGER, 1974. Observations on bovine parturition. Proc. Aust. Soc. Anim. Prod. 10:314-317.

HAFEZ, E. S. E., and J. A. LINEWEAVER, 1968. Suckling behaviour in natural and artificially fed neonate calves. Z. Tierpsychol. 25:187-198.

JONES, J. E. T., 1966. Parturition in the sow. I. The pre-partum phase. Br. Vet. J. 122:420-426.

LECCE, J. G., 1975. Rearing piglets artificially in farm environment: A promise unfulfilled. J. Anim. Sci. 41:659-666.

LIGGINS, G. C., 1968. Premature parturition after infusion of corticotrophin or cortisol into foetal lambs. J. Endocrinol. 42:323-329.

——, S. A. GRIEVES, J. Z. KENDALL, and B. S. KNOX, 1972. The physiological roles of progesterone, oestradiol-17β and prostaglandin $F_{2\alpha}$ in the control of ovine parturition. J. Reprod. Fertil., Suppl. 16:85-103.

LIGGINS, G. C., P. C. KENNEDY, and L. W. HOLM, 1967. Failure of initiation of parturition after electrocoagulation of the pituitary of the foetal lamb. Am. J. Obstet. Gynecol. 98:1080-1086.

MORGAN, P. D., and G. W. ARNOLD, 1974. Behavioural relationships between Merino ewes and lambs during the four weeks after birth. Anim. Prod. 19:196-206.

MUNRO, J., 1956. Observations on the suckling behaviour of young lambs. Br. J. Anim. Behav. 4:34-36.

NOLL, M. T., 1979. Use of cow colostrum in artificial rearing of baby pigs. M.S. Thesis, Kansas State University Library, Manhattan, Kans.

OWEN, F. G., 1978. The calf: birth to 16 weeks. In C. J. Wilcox (Ed.), Large Dairy Herd Management. University Presses of Florida, Gainesville, Fla.

POMEROY, R. W., 1960. Infertility and neonatal mortality in the sow. III. Neonatal mortality and fetal development. J. Agric. Sci. 54:31-56.

RANDALL, G. C. B., 1972. Observations on parturition in the sow. I. Factors associated with the delivery of the piglets and their subsequent behaviour. Vet. Rec. 90:178-182.

ROSSDALE, P. D., 1967. Clinical studies on the newborn Thoroughbred foal. Br. Vet. J. 123:470-481.

——, and R. V. SHORT, 1967. The time of foaling of Thoroughbred mares. J. Reprod. Fertil. 13:341-343.

SCHLOETH, R., 1961. Das Sozialleben des Camargue-rindes. Qualitative und quantitative Untersuchungen über die sozialen Beziehungen—insbesondere die soziale Rangordnung—des halbwilden französischen Kampfrindes. Z. Tierpsychol. 18:574-627.

SELMAN, I. E., A. D. McEWAN, and E. W. FISHER, 1970a. Studies on natural suckling in cattle during the first eight hours post partum. I. Behavioural studies (dams). Anim. Behav. 18:276-283.

——, 1970b. Studies on natural suckling in cattle during the first eight hours post partum. II. Behavioural studies (calves). Anim. Behav. 18:284-289.

——, 1971. Absorption of immune lactoglobulin by newborn dairy calves. Res. Vet. Sci. 12:205-210.

SHORT, R. V., 1960. Blood progesterone levels in relation to parturition. J. Reprod. Fertil. 1:61-70.

THORBURN, G. D., D. H. NICOL, J. M. BASSETT, D. A. SHUTT, and R. I. COX, 1972. Parturition in the goat and sheep: changes in corticosteroids, progesterone, oestrogens and prostaglandin F. J. Reprod. Fertil., Suppl. 16:61-84.

TYLER, S. J., 1972. The behaviour and social organization of the New Forest ponies. Anim. Behav. Monogr. 5:87-196.

WALKER, D. M., 1950. Observations on behaviour in young calves. Bull. Anim. Behav. 1:5-10.

WITHERS, F. W., 1953. Mortality rates and disease incidence in calves in relation to feeding, management and other environmental factors. VI. Br. Vet. J. 109:122-131.

Author Index

Subject Index